Ship Handling

Theory and practice

Other Works Published by D. J. House

Seamanship Techniques, 3rd Edition, 2004, Elsevier Ltd., ISBN 0750663154 (first published in two volumes 1987)

Seamanship Techniques, Volume III 'The Command Companion', 2000, Elsevier Ltd., ISBN 0750644435

Marine Survival and Rescue Systems, 2nd Edition, 1997, Witherby, ISBN 1856091279

An Introduction to Helicopter Operations at Sea – A Guide for Industry, 2nd Edition, 1998, ISBN 1856091686

Cargo Work, 7th Edition, 2005, Elsevier Ltd., ISBN 0750665556

Anchor Practice – A Guide for Industry, 2001, Witherby, ISBN 1856092127

Marine Ferry Transports – An Operators Guide, 2002, Witherby, ISBN 1856092313

Dry Docking and Shipboard Maintenance, 2003, Witherby, ISBN 1856092453

Heavy Lift and Rigging, 2005, Brown Son and Ferguson, ISBN 085174 720 5

The Seamanship Examiner, 2005, Elsevier Ltd., ISBN 075066701X

Navigation for Masters, 3rd Edition, 2006, Witherby, ISBN 1865092712

Website www.djhouseonline.com

Ship Handling

Theory and practice

D.J. House

LONDON AND NEW YORK

First published 2007 by Butterworth-Heinemann

This edition published 2012 by Routledge
2 Park Square, Milton Park, Abingdon, Oxon, OX14 4RN
711 Third Avenue, New York, NY 10017

Routledge is an imprint of the Taylor & Francis Group, an informa business

First Edition 2007

Notice
No responsibility is assumed by the publisher for any injury and/or damage to persons or property as a matter of products liability, negligence or otherwise, or from any use or operation of any methods, products, instructions or ideas contained in the material herein. Because of so many variable factors involved in the practice of ship handling, the publisher and author cannot be held in any way responsible for associated industrial practice as described within this publication

Repeated use of 'he or she' can be cumbersome in continuous text. For simplicity, therefore, the male pronoun predominates throughout this book. No bias is intended, as the position of an Officer, Chief Mate, Helmsman, Engineer, etc. can equally apply to a female worker.

British Library Cataloguing in Publication Data
A catalogue record for this book is available from the British Library

Library of Congress Cataloguing in Publication Data
A catalogue record for this book is available from the Library of Congress

ISBN: 978-0-7506-8530-6

Typeset by Charon Tec (A Macmillan Company), Chennai, India
www.charontec.com

I would like to express my thanks and appreciation to Mr. John Finch, Master Mariner, Lecturer Nautical Studies, who has provided guidance and support on this particular publication and all the author's previous works.

It has been a privilege to receive his constructive and honest criticism over the many years we have been friends.

Contents

About the author

David House is currently engaged in the writing and the teaching of maritime subjects, with his main disciplines being in the Seamanship and Navigation topics. Following a varied seagoing career in the British Mercantile Marine, he began a teaching career at the Fleetwood Nautical College in 1978. He also commenced writing at about this time and was first published in 1987 with the highly successful "Seamanship Techniques" now in its 3rd edition and distributed worldwide.

Since this initial work, originally published as two volumes, he has written and published fourteen additional works on a variety of topics, including: Heavy Lifting Operations, Helicopter Operations at Sea, Anchor Work, Drydocking, Navigation for Masters, Cargo Work, Marine Survival and Ferry Transport Operations.

This latest publication is designed as a training manual, to highlight the theory and practice of ship handling procedures, relevant to both the serving operational officer as well as the marine student. It encompasses the experiences of the author in many of the scenarios and reflects on the hardware employed in the manoeuvring and the control of modern shipping today.

Preface

The reality of handling the ship is a world apart from the theory. No publication can encompass the elements of weather and features of water conditions to make the practice and theory one and the same. The best any book can hope for is to update the mariner with the developments in hardware employed to effect modern-day manoeuvres. Since the demise of sail, machinery and manoeuvring aids have continued to improve and provide additional resources to the benefit of Masters, Pilots and others, charged with the task of handling both large and small power-driven vessels.

Maritime authorities are united in establishing a safe and pollution-free environment. Internationally, it is these interests that provide the desired protection for operators to conduct their trade in some of the most active and busiest areas of the world. The theory of a manoeuvre may be ideally suited for a certain port at a certain time, but the many variables involved may make the same manoeuvre totally unsuitable at another time. Ship handlers and controllers must therefore be familiar with the capabilities of the ship, while at the same time be flexible in the use of resources against stronger currents or increased wind conditions.

Knowing what to do and when to do it: in order to attain the objective is only half of the task. The reasoning behind the actions of the ship handler will tend to be based on the associated theory at the root of any handling operation. Such knowledge – coupled with main engine power and steering, anchors and moorings, tugs and thrusters, if fitted – can be gainfully employed to achieve a successful docking or unberthing.

Practice with different ships, and fitted with different manoeuvring aids, tends to increase the experience of the would-be ship handler. Training for junior officers to increase their expertise in the subject is unfortunately extremely limited. Unless Ship's Masters allow 'hands on' accessibility, few have the early opportunity to go face to face with a subject which is not an exact science. The theoretical preparation, the advance planning and the execution of any manoeuvre will not materialise overnight. And an understanding of the meteorological conditions may not initially be seen as a relevant topic, but ship handling against strong winds with a high freeboard vessel is somewhat different to manoeuvring with a large fully loaded tanker with reduced freeboard in calm sea conditions.

The purpose of the text, therefore, is to combine the hardware, with the theory in variable weather and operating conditions. Ship handling is not a stand alone topic and, by necessity, must take account of the many facets affecting a successful outcome. Knowing the theory is necessary, putting it into practice is essential.

David J. House

Acknowledgements

I would like to express my thanks and gratitude to the following companies and individuals who have kindly contributed to this publication:

Becker Rudder KSR
B + V Industrietechnik
Dubia Dry Docks, U.A.E.
Holland Roer-Propeller Propulsion Systems and Bowthrusters
Smit Maritime Contractors, Europe and Smit International
Stena Line Ferries (Ex., P & O Ferries Dover)
MJP Waterjets

Technical content advisor:

Mr. J. Finch Master Mariner, Senior Lecturer Nautical Studies
I.T. Consultant:
Mr. C.D. House

Additional photography:

Mr. Stuart Mooney, Chief Officer (MN) Master Mariner
Mr. Paul Brooks, Chief Officer (MN) Master Mariner
Mr. John Legge, Chief Officer (MN) Master Mariner
Mr. K.B. Millar, Master Mariner. Lecturer, Nautical Studies
Mr. Mathew Crofts, Master Mariner. Lecturer, Nautical Studies
Mr. J. Warren, Master Mariner. Lecturer, Nautical Studies
Mr. J. Leyland, Lecturer, Nautical Studies
Mr. N. Sunderland, Chief Officer (MN)
Miss Martel Fursden, 2nd Officer (MN)

Additional computer artwork:

Mr. F. Saeed, Master Mariner. Lecturer Nautical Studies MSc

Meteorological tables common to the marine environment

Fog and visibility table

Scale number	*Description and range*
0	Dense fog, targets not visible at 50 metres
1	Thick fog, targets not visible at 1 cable
2	Fog, targets not visible at 2 cables
3	Moderate fog, targets not visible at 0.5 mile
4	Mist or haze, targets not visible at 1 n/mile
5	Poor visibility, targets not visible at 2 n/miles
6	Moderate visibility, targets not visible beyond 5 n/miles
7	Good visibility, targets visible up to 10 n/miles
8	Very good visibility, targets visible up to 30 n/miles
9	Excellent visibility, targets visible beyond 30 n/miles

Sea state table

Descriptive state of sea waves	*Wave height in metres*
Calm – glassy	0
Calm – ripples	0–0.1
Smooth wavelets	0.1–0.5
Slight	0.5–1.25
Moderate	1.25–2.5
Rough	2.5–4.0
Very rough	4.0–6.0
High	6.0–9.0
Very high	9.0–14.0
Phenomenal	Over 14 metres high

Swell

Length of swell	*Length in metres*
Short	0 to 100
Average	100 to 200
Long	Over 200
Height of swell	*Height in metres*
Low	0 to 2.0
Moderate	2.0 to 4.0
Heavy	Over 4.0

The Beaufort Wind Scale

Force	*Description*	*Sea state*	*Speed in knots*
0	Calm	Smooth	0–1
1	Light airs	Small wavelets	1–3
2	Slight breeze	Short waves, cresting	4–6
3	Gentle breeze	Small waves, breaking	7–10
4	Moderate breeze	Definite whitecaps	11–16
5	Fresh breeze	Moderate waves	17–21
6	Strong breeze	Larger waves	22–27
7	Moderate gale	Spindrift formed	28–33
8	Fresh gale	Much spindrift	34–40
9	Strong gale	Seas start to roll	41–47
10	Whole gale	Seas roll and break heavily	48–55
11	Storm	Surface all white big seas	56–65
12	Hurricane	Enormous seas	Above 65

Weather notations and symbols as plotted on synoptic weather charts

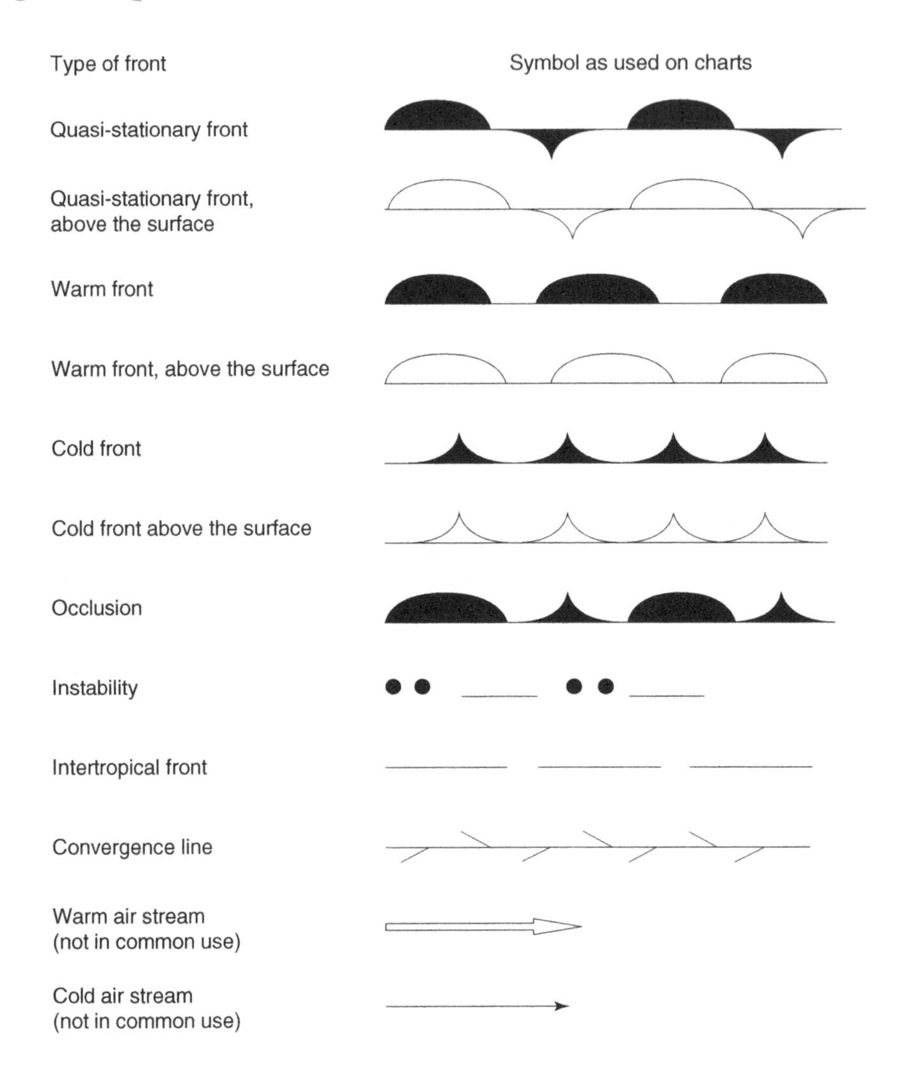

List of abbreviations associated with ship handling and shipboard manoeuvres

AC	Admiralty Cast (Class)
ACV	Air Cushion Vessel
AHV	Anchor Handling Vessel
AIS	Automatic Identification System
AKD	Auto Kick Down
AM	Admiralty Mooring
AMD	Advanced Multi-Hull Design
AMVER	Automated Mutual Vessel Rescue system
ARPA	Automatic Radar Plotting Aid
ASD	Azimuth Stern Drive
ATT	Admiralty Tide Tables
AUSREP	Australian Ship Reporting system
BS	Breaking Strength
CBD	Constrained by Draught
CD	Chart Datum
CG	Coast Guard
CMG	Course Made Good
CNIS	Channel Navigation Information Service
C/O	Chief Officer
COLREGS	The Regulations for the Prevention of Collision at Sea
CPA	Closest Point of Approach
CPP	Controllable Pitch Propeller
CQR	Chatham Quick Release (anchor type) (doubtful)
CSP	Commencement of Search Pattern
CSWP	Code of Safe Working Practice
D	Depth
DAT	Double Acting Tanker
DB	Double Bottom (tanks)
DC	Direct Current
DGPS	Differential Global Positioning System

DNV-W1 One Man Bridge Operation (DNV requirement)
DP Dynamic Positioning
DR Dead Reckoning
DSC Dynamically Supported Craft (Hydrofoils)
DSV Diving Support Vessel
DV Desired Value
DWA Dock Water Allowance
DWT (dwt) Deadweight

ECDIS Electronic Chart Display and Information System
ECR Engine Control Room
ENC Electronic Navigation Chart
ETA Estimated Time of Arrival
ETD Estimated Time of Departure
ETV Emergency Towing Vessel

FFTS Flat Fluke Twin Shank
FMECA Failure Mode Effective Critical Analysis
FPSO Floating Production Storage Offloading system
FRC Fast Rescue Craft
FSE Free Surface Effect
FSU Floating Storage Unit
FW Fresh Water
FWE Finished With Engines

G Representative of the Ship's Centre of Gravity
GM Metacentric Height
GPS Global Positioning System
Grt Gross registered tons
GT Gas Turbine

HFO Heavy Fuel Oil
h.p. Horse power
HSC High Speed Craft
HW High Water

IACS International Association of Classification Societies
IALA International Association of Lighthouse Authorities
IAMSAR International Aeronautical and Maritime Search and Rescue manual
IIP International Ice Patrol
IMO International Maritime Organization
INS Integrated Navigation System
IPS Integrated Power System (Controllable Podded Propulsion Units)
IWS In Water Survey

Kg Kilograms
Kts Knots
kW Kilowatt

LAT	Lowest Astronomical Tide
LBP	Length Between Perpendiculars
LCD	Liquid Crystal Display
LHC	Left Hand Controllable
LHF	Left Hand Fixed, propeller
LMC	Lloyds Machinery Certificate
LOA	Length Overall
LSA	Life Saving Appliances
LW	Low Water
M	Representative of the Ship's Metacentre
M	Metres
MAIB	Marine Accident Investigation Branch
MCA	Maritime and Coastguard Agency
MCTC	Moment to Change Trim 1 Centimetre
Medivac	Medical Evacuation
MGN	Marine Guidance Notice
MHWN	Mean High Water Neaps
MHWS	Mean High Water Springs
MLWN	Mean Low Water Neaps
MLWS	Mean Low Water Springs
MMSI	Maritime Mobile Service Identity Number
mm	millimetres
MoB	Man overboard
MPCU	Marine Pollution Control Unit
MRCC	Marine Rescue Co-ordination Centre
MSC	Maritime Safety Committee (of IMO)
MSI	Marine Safety Information
MSN	Merchant Shipping Notice
MV (i)	Motor Vessel
MV (ii)	Measured Value
nm	nautical mile
NUC	Not Under Command
NVE	Night Vision Equipment
OiC	Officer in Charge
OIM	Offshore Installation Manager
OMBO	One Man Bridge Operation
OOW	Officer Of the Watch
O/S	Offshore
OSC	On Scene Co-ordinator
PEC	Pilot Exemption Certificate
PSC	Port State Control
RAF	Royal Air Force
RHC	Right Hand Controllable

RHF Right Hand Fixed, propeller
RMS Royal Mail Ship
RN Royal Navy
RoPax Roll on–Roll off Passenger Vessel
Ro–Ro Roll on–Roll off
RoT Rate of Turn
RPM Revolutions Per Minute

SAR Search and Rescue
SBE Stand By Engines
SBM Single Buoy Mooring
s.h.p. Shaft Horse Power
SMC SAR Mission Controller
SMG Speed Made Good
SPM Single Point Mooring
SQ Special Quality
SS Steam Ship
Stb'd Starboard
SW Salt Water
SWATH Small Waterplane Area Twin Hull
SWL Safe Working Load

TMC Transmitting Magnetic Compass
TRS Tropical Revolving Storm
TSS Traffic Separation Scheme
TVF Tip-Vortex – Free

UKC Under Keel Clearance
ULCC Ultra Large Crude Carrier
UMS Unmanned Machinery Space
USCG United States Coast Guard

VCR Voith Cycloidal Rudder
VDR Voyage Data Recorder
VHF Very High Frequency
VLCC Very Large Crude Carrier
VLGC Very Large Gas Carrier
VSP Voith Schneider Propeller
VTMS Vessel Traffic Management System
VTS Vessel Traffic System

WBT Water Ballast Tank
WiG Wing in Ground effect
W/L Water line
WPC Wave Piercing Catamaran

Definitions, terminology and shipboard phrases relevant to the topic of ship handling and this text

Advance Described by that distance a vessel will continue to travel ahead on her original course while engaged in a turning manoeuvre. It is measured from that point at which the rudder is placed hard over, to when the vessel arrives on a new course 90° from the original.

Air Draught That measurement from the waterline to the highest point of the vessel above the waterline.

Anchorage A geographic area suitable for ships to lay at anchor. Ideally, it would have good holding ground and be free of strong currents and sheltered from the prevailing weather. It is usually identified on the nautical chart by a small blue anchor symbol.

Anchor Aweigh An expression used to describe when the vessel breaks the ground and no longer secures the vessel. The cable is in the up/down position and the vessel is no longer attached to the shore by the anchor.

Anchor Ball A round ball shape, black in colour, which is required to be shown by vessels at anchor, under the Regulations for the Prevention of Collision at Sea.

Anchor Bearings Those bearings taken to ascertain the ship's position when she has become an anchored vessel.

Anchor Buoy An identification buoy used to denote the position of the deployed anchor. It is hardly ever used by commercial shipping in this day and age.

Anchor Coming Home The action of drawing the anchor towards the ship as opposed to pulling the ship towards the anchor.

Anchor Plan A preparatory plan made by the Master and ship's officers prior to taking the ship to an anchorage.

Anchor Warp A steel wire hawser length, usually attached to a short length of anchor chain or directly onto the anchor for warping the vessel ahead or astern.

Astern (i) The movement of the ship's engines in reverse, to cause the stern first movement of the vessel; (ii) Descriptive term used to describe an area abaft the ship's beam and outside of the vessel's hull.

Auto-Pilot A navigation bridge control unit employed to steer the vessel in an unmanned mode. Various controls can be input by the operator to compensate for sea and weather conditions but the unit is effectively a free-standing steering unit.

AziPod Trade name for a rotable thruster unit with or without ducting, turning through 360° rotation and providing propeller thrust in any direction.

Baltic Moor A combination mooring of a vessel alongside the berth which employs a stern mooring shackled to the offshore anchor cable in the region of the 'ganger length'. When approaching the berth, the offshore anchor is deployed and the weight on the cable and the stern mooring act to hold the vessel just off the quay.

Band Brake A common type of brake system found employed on windlasses. The band brake is a screw on friction brake, designed to check and hold the cable lifter (gypsy) when veering anchor cable.

Beaching The term used to describe the act of the ship taking the ground intentionally. It is a considered action if the ship is damaged and in danger of being lost.

Bight The middle part of a line or mooring. It may be seen as a loop in a rope or may be deliberately created to run around a bollard providing two parts of a mooring (instead of one). It is considered extremely dangerous to stand in the bight of a rope and persons in charge of mooring decks should watch out for the young or less experienced seafarers, when working with rope bights.

Bitter End That bare end of the anchor cable which is secured on a quick release system at the cable locker position.

Bitts A seaman's term for describing the ship's bollards.

Bollard Pull An expression which is used in charter parties to grade the capacity of a tug and its efficiency. The bollard pull is assessed by measurement, against the pulling capacity of a tug, as measured by a dynamometer. The thrust, or force developed is known as 'Bollard Pull' and is expressed in tonnes. It is useful for marine pilots to assess the wind force affecting the ship against the available 'bollard pull'.

Bow Anchor A vessel is normally fitted out with two working bow anchors. Specialist vessels may also be equipped with additional anchors for specific trade or operations, i.e. stern anchor.

Bow Stopper A collective name to describe either a guillotine or a compressor. Both of which act as an anchor cable stopper. It is one of the securing devices applied to the anchor cable when the vessel is at sea. Alternatives: the AKD stopper (Auto Kick Down).

Breakers These are waves which break against the shoreline producing surf.

Breast Line A ship's mooring line which is stretched at right angles to the fore and aft line of the vessel. By necessity, they are generally short compared to the long drift of head or stern lines, the function of the breast line being to retain the vessel alongside the quay.

Brought Up An expression used to describe when the vessel is 'Brought Up' to the anchor, when the anchor is deployed and holding. The scope of cable is observed to rise and fall back in a catenary indicating that the vessel is riding to her anchor and not dragging her anchor.

Bruce Anchor A trade name to describe a specialist anchor manufactured by the anchor company 'Bruce Ltd'. The original 'Bruce' design incorporated the hook effect of the Admiralty Pattern Anchor and the Spade effect of the stockless anchor to produce a high holding power anchor with no moving parts.

Bullring Often referred to as a centre lead, set well forward in the eyes of the vessel. It is often employed for towing or accommodating buoy mooring lines. When not employed with moorings it is often used to hold a company or ship's emblem.

Cable A nautical measurement equivalent to one tenth of a nautical mile, or 100 fathoms (also 608 feet).

Cable Holder A cable lifter which is mounted horizontally as opposed to vertically on a windlass axle. Some passenger and warship vessels operate anchors with cable holders rather than windlass operations.

Caisson The term used to describe a dry dock or dock gate system.

Capstan A vertically mounted warping drum with its motor secured below decks. The sides of the drum are fitted with 'whelps' to provide improved holding for mooring rope turns.

Carry Up A term used to refer to moorings being carried up the quayside when mooring alongside or entering a dock, the moorings usually then being employed to warp the vessel ahead or astern or assist in the manoeuvring of the vessel.

Cavitation A physical phenomena experienced in the region of a rotating propeller and its supporting structure. The cause is generally an air bubble flow which is non-uniform, associated with the water flow from the propeller action. Extensive cavitation effect can give rise to excessive corrosion in the propeller area of the vessel.

Chart Datum A plane of reference for charted depths. The United Kingdom employs the lowest astronomical tide, the lowest water prediction. In the United States, it is the mean low water.

Circle of Swing That area that a vessel will swing over when lying to an anchor. The circle of swing can be reduced by mooring to two anchors.

Coir Springs Heavy duty harbour moorings manufactured in coir rope. They are designed to be picked up by a vessel mooring in a harbour, usually where heavy swells are experienced. Commonly referred to as 'storm moorings'. Common to ports on the Pacific rim, they are used in addition to the ship's own moorings.

Composite Towline A towline which is established by employing the ship's anchor cable secured to the towing vessel's towing spring.

Con (Conn) An expression used to describe the person who has the control of the navigation of the vessel.

Contra-rotating Propellers Two propellers mounted on the same shaft rotating in opposite directions to balance torque.

Controllable Pitch Propeller A propeller which is constructed in such a manner that the angle of the blades can be altered to give a variable pitch angle. Namely from zero pitch to maximum pitch ahead or astern.

Crest of a Wave The peak or highest point of a wave. Opposite to the trough of a wave.

Cross A term used to describe a 'foul hawse' where the anchor cables have crossed over as the vessel has swung through 180°.

Devils Claw A securing device used to secure the anchor cable, when the vessel is at sea.

Docking Winch The name given to the aft mooring deck winch which is employed for use with the stern mooring lines. It may also have an integrated cable lifting operation if the vessel is equipped with a stern anchor.

Double-up When referred to moorings, means the act of doubling a single part mooring to a double mooring, e.g. double up the forward spring line.

Drag An effect which opposes the ship's forward motion and can be caused by shell/hull friction, rudder action or appendages extending from the hull, effectively reducing the ship's speed. The term is also used to describe a ship dragging its anchor.

Dragging Anchor An expression used to describe a vessel which is moving over the ground when its anchor is not dug in and holding.

Draught The depth measure of a freely floating ship. It is the vertical measurement between the keel of the ship to the waterline (alternative spelling 'draft').

Drawing the Anchor Home A phrase which describes pulling the anchor home towards the ship as opposed to pulling the ship towards the anchor.

Dredging (an anchor) A term when used in conjunction with an anchor, it means the deliberate dragging of an anchor when at short stay, over the ground of the sea bed.

Drop an Anchor Underfoot The action of letting go a second anchor at short stay. It is usually done to reduce the 'Yaw' or movement by the ship about the riding cable. It tends to act as a steadying influence to oscillations by the ship when at a single anchor.

Ducting A term used to describe the propeller being encompassed by a partial steel tunnel to 'chunnel' the water flow more directly onto the propeller blades.

Ebb Tide The tidal flow of water out of a port or harbour away from the land.

Elbow A term used to describe a 'foul hawse' where both the deployed anchor cables have crossed over and the vessel has turned 360°.

Even Keel An expression which describes a vessel which is without any angle of list, is said to be on 'even keel'.

Fairway That navigable and safe area of a harbour approach which may include the main shipping channel. It is usually marked with a fairway buoy.

Falling Tide A term used to describe when the tide is falling on the ebb and the depth of water is decreasing.

Fender A purpose-built addition to the ship's hull to prevent damage to the hull when landing alongside a jetty or other hard surface. It may also be a portable device suspended on a lanyard to protect the hull from damage when strategically placed between the quayside and the ships hull to cushion and protect the ships side.

Fetch Described as the distance that the wind blows over the sea without encountering any appreciable interference from land masses. The term was also previously used in sailing vessels, i.e. to 'fetch up on a starboard tack'.

Final Diameter Is defined as that internal diameter of the ships turning circle where no allowance has been made for the decreasing curvature as experienced with the tactical diameter.

Fine A bearing reference which indicates an observation bearing, less than ½ compass point off the bow, but not dead ahead.

Flipper Delta Anchor A modern high holding power anchor which can have the angle of the flukes pre-set at a variable, desired angle, prior to deployment.

Flood Tide A tide which flows into a port or harbour or into and towards the land. Opposite to an ebb tide which represents the tide flowing outwards.

Fog (see Visibility Table) Is a condition formed when cloud occurs at ground (sea) level. There are two recognized forms, namely radiation fog and advection fog. In all cases, visibility is impaired to less than 1000 metres. When mixed with polluted air it is termed as smog.

Foul Anchor A description given to an anchor which is obstructed by a foreign object (usually from the sea bed) or fouled by its own anchor cable. It is only usually detected when the anchor is heaved up to be stowed.

Foul Hawse An expression which describes when both anchor cables have become entwined with each other. It can occur when two anchors are deployed at the same time, as in a running moor. A change in the wind direction, left unobserved, causes the vessel to swing through the line of cables causing the foul.

Ganger Length A short length of anchor cable set between the anchor crown 'D' shackle and the first joining shackle of the cable. The length may consist of just a few links which may or may not contain a swivel fitting.

Girding a Tug The action of pulling on a towline at right angles to the fore and aft line of the tug, in a manner likely to cause a capsize motion on the tug. Alternative term is 'girting'.

Gob Rope (Alt., Gog Rope) A strong rope (or wire plus heavy shackle) set over the tow line of a tug. Its function is to bowse the towline down towards the aft end of the tug, so changing the direction of weight on the tug. Its function is also to reduce the risk of the tug being girted and caused to capsize.

Grounding A term used to describe when a ship touches the sea bottom accidentally. It occurs generally through poor navigation and the lack of underkeel clearance. The severity of any damage incurred will depend on the speed of striking and the nature of the ground that the vessel contacts.

Hang off an Anchor The operation of detaching the anchor from its cable and hanging it off, usually at the break of the forecastle. The operation is carried out when the vessel needs to moor up to mooring buoys by its anchor cable or if it is expecting to be towed by means of a composite towline.

Hawser A term which refers to a mooring line in the United Kingdom, meaning a large diameter fibre rope or wire rope.

Heading That direction in which the ship is pointed. It is usually compass referenced.

Headreach That distance that the vessel will move ahead after the engines have been stopped and before the ship stops steering.

Headway The forward movement of the vessel through the water. Opposite to sternway, when the vessel is moving astern.

Head Wind A condition when the wind is from the opposite direction to the ships course. Similar meaning for a head sea.

Heaving Line A light line fitted with a weighted end (Monkey fist) which can be thrown from ship to quay or quay to ship (depending on wind direction). It is used for the connection and passing of heavy moorings between the deck crew and the wharf men.

Heave To A reduction of the ship's speed, usually made in heavy weather conditions. The speed reduction is reduced to maintain steerage and hold the ship's head into the prevailing weather and sea direction.

Heel That angular measure that a vessel will be inclined by an external force, e.g. wind or waves. The condition can also occur during a turning manoeuvre.

Helm A term which refers to the tiller or ship's steering wheel. A vessel may carry 'helm' as in having a turn of the ship's wheel held to retain the vessel on course. It is also the name given to one of the controlling elements of automatic steering units.

Helmsman Alternative name for a quartermaster, who steers the ship to the orders of the watch officer, master or pilot.

Holding Ground A description of the type of ground into which a vessel is letting go her anchor, e.g. mud, sand, broken shell, etc. There is good holding ground for the anchor and bad holding ground for the anchor.

Holding Power An expression used to describe the holding power of an anchor. Some anchors like the ' Bruce' or the 'AC14' are recognized as having High Holding Power qualities, much more than a conventional anchor design like the stockless.

Hove in Sight An expression which refers to heaving the anchor clear of the water surface. Once the anchor is sighted, the bridge should be informed it is sighted and clear.

Hydro-Lift A dry docking system which employs hydrostatic force to lift and lower vessels to be docked. The system operates similar for vessels which move through the locking system of the 'Panama Canal'. The most well known example is at Lisnave, in Portugal, where a wet basin allows three large vessels to be docked at the same time.

Interaction A term which describes the behaviour of a ship when it is influenced by either a fixed object like the proximity of the land or another vessel passing too close. There are several types of interaction (see squat) all of which are undesirable and tend to cause movement of the vessel outside the influences of the controller.

Joining Shackle A single specialized shackle that joins two shackle lengths of cable. The most common joining shackle employed is the 'kenter shackle' but 'D' lugged joining shackles are also employed for the same purpose.

Jury A term meaning temporary or improvised. As with a 'jury rudder'.

Kedge The forced movement of a vessel astern by the laying of a 'kedge anchor' to pull the vessel astern, usually off a bank. Some ships carry a specific kedge anchor but the practice of carrying this, is now rare.

Knot The nautical unit of speed which equates to approximately 1⅐*th* of a statute mile per hour (One knot = one nautical mile per hour).

Kort Nozzle Trade name for an encased propeller which is capable of rotating through 360°. Extensively used in tugs.

Landlocked When a vessel is surrounded by land as in a bay or other restricted waters she is said to be landlocked.

Lanyard A short line used to hold or secure something, i.e. a bucket or a sidearm. Previously used in sailing ships' rove through a block to tighten rigging.

Lead A narrow, navigable channel through an ice field.

Lee That side of the ship that lies away from the wind. Opposite to the weather side.

Lee Shore A land mass or coastline towards which the wind is blowing. A loss in engines off a lee shore could lead to the vessel being blown aground.

Leeward Refers to that side which is away from the wind. It is pronounced 'lu-ward' and is the side opposite to windward.

Leeway That sideways movement of a vessel away from the designated course due to the force of the wind.

Let Go An expression which describes the release of the anchor from the windlass braking system. With the advent of heavier anchors being installed on larger vessels fewer ships are actually 'letting go anchors'. The modern tendency is to 'walk back' the anchor cable under full control.

Log (i) A device for measuring the ship's mileage and subsequently its speed;
(ii) Shortened term for the ship's logbook.

Long Stay An expression which describes the line of cable when the vessel rides to an anchor; the line of cable being observed as a line near parallel to the water surface. Compared to short stay, where the angle of cable is at an acute angle to the water surface.

Lubber Line A reference mark usually found on the inside of the compass bowl in line with the ship's head. Employed with the steering of the vessel.

Magnetic Compass A ship's compass which aligns to the magnetic North Pole. It is considered the most important instrument on the vessel as it does not rely on an external power source like the gyroscopic compass.

Mediterranean Moor A ship's mooring which allows the vessel to be secured to the quay by stern moorings while the bow is held fixed by deploying both bow anchors. The mooring is suitable for non-tidal waters, like the Mediterranean Sea.

Messenger Line A light line employed as an easy to handle length, used to pass a heavy mooring hawser, as with a 'slip wire'.

Monkey Fist A heavy knot made at the end of a heaving line to provide a weighted end to improve throwing.

Mooring (i) The term used to describe a vessel secured with two anchors;
(ii) The term used to describe a vessel which is being tied up to the quayside or moored to buoys.

Mooring Anchor A heavy anchor employed as a permanent mooring for buoys or, in some cases, offshore installations.

Mooring Boat A small boat employed to carry ship's moorings to the shore or to mooring buoys. It is usually manned by a minimum of two men, one of which may have to 'jump the buoy' when securing to buoys.

Mooring Buoy A large buoy to which ships can moor using mooring lines or by means of the anchor cable once the anchor has been 'hung off'.

Mooring Deck That area of a ship from which the moorings are run ashore and secured. The vessel would normally have a forward mooring deck and an aft mooring deck. The forward deck usually accommodates the anchor arrangement.

Mooring Line A natural fibre or manmade fibre rope used to tie up and secure the vessel to quaysides or buoys. A generic term which can also include mooring wires.

Mooring Shackle A heavy duty bow shackle, listed under the anchors and cables accessories. It is used when the vessel needs to moor up to buoys.

Mooring Swivel An additional fitting placed into the anchor cable when mooring to buoys or to two anchors for a lengthy period of time. The swivel ensures that the cable does not become fouled and twisted as the vessel turns on the mooring.

Mushroom Anchor A type of mooring anchor so-called because of its shape being similar to a mushroom. It is used extensively as a permanent mooring for navigation marks and buoys.

Neap Tide A tide which occurs twice a month of reduced range or velocity. It occurs when the moon is in quadrature with the sun (opposite to a spring tide).

Not Under Command The term given to a vessel which is unable to manoeuvre as required by the 'Rules of the Road' because of exceptional circumstances.

Officer Of the Watch (OOW) The description of the navigation officer who is placed in charge of the watch at sea. The OOW is responsible for the safe navigation of the vessel during his or her period of duty and is expected to have full control of the ship's course, speed and navigation aids.

Offshore Wind A direction of wind which blows towards the sea away from the land.

Old Man The term used to describe a single roller lead, mounted on a pedestal. It is often used to change the direction of a mooring line away or towards the lead of the windlass.

Onshore That direction towards the coastline from seaward (opposite is offshore).

Open Moor The name given to a mooring which employs two anchors, each one deployed about 20° off each bow. The mooring is used in non-tidal waters to provide additional holding power against a strong flowing stream.

Overhauling A term used to describe one vessel overtaking and passing another when both vessels are going in the same direction. *NB*. Can also mean a term in maintenance to overhaul a ship or piece of machinery.

Panama Lead Often referred to as a pipe lead which prevents moorings from accidentally jumping out of the lead when under weight. For this reason, many seamen prefer the use of panama leads as opposed to roller leads.

Period of Encounter May be considered as the period of time between the passage of two successive wave crests to pass a fixed point, namely the position of the ship.

Period of Pitch Is defined by that time the bows of a ship start to make a rise from the horizontal, then fall back below the horizontal and then return to it.

Period of Roll Defined by that time period a vessel will roll from one side to the other and return, when rolling freely.

Pitch (i) The vertical upward and downward movement of the vessel along its fore and aft line caused by head or following seas; (ii) That angle a propeller blade will make with a perpendicular plane of the axis of the propeller. The pitch angle will vary along the length of the blade. Propeller pitch can also be expressed as the distance the propeller will move forward in one revolution through a soft medium (e.g. water).

Pivot Point That position aboard the vessel about which the ship rotates when turning. In conventional vessels the 'pivot point' was approximately one third (⅓) of the ship's length, measured from forward, when moving ahead. The position of the pivot point will change when going astern and with the types of ship construction.

Plimsoll Mark The loadline markings painted on the ship's side to indicate the maximum load draught that the vessel may load her cargo under different conditions.

Plummer Block An alignment support bearing, for the ship's propeller shaft.

Pointing Ship The action of changing the ship's head when lying to a single anchor. It is achieved by passing a stern mooring wire forward to secure to the anchor cable. The cable is then veered causing the vessel to lay at an acute angle to the flow. It is employed to create a 'lee' if working small craft to clear from the weather side.

Poop Deck A term which refers to the aftermost deck of the vessel. It usually carries a superstructure known just as the 'poop'. Originally it developed from what was known as the 'aft castle' of medieval sailing ships and was later to provide additional buoyancy to the ship as well as accommodation for the Master and Officers.

Pooped A term which describes a large sea or wave which breaks over the poop deck area when the vessel is running with a following sea.

Port A reference to the left side of the vessel when looking forward.

Pounding A term which describes the heavy contact of the ship's fore part when pitching in a seaway. This is a violent contact and may cause ship damage, it is sometimes referred to as slamming. The effect of pounding can usually be tempered by a reduction in speed.

Propeller Diameter That diameter described by the tips of the propeller blades when turning.

Propeller Ducting A cylindrical steel casing set around the propeller, often fitted with reaction vanes to concentrate the flow of water directly to the turning area of the propeller. Also known as a 'Propeller Shroud' keeping the wash from the propeller into a confined area. Popular with smaller craft and harbour authorities because they tend to reduce erosion of river and canal banks.

Propeller Pitch Described by the axial distance moved forward by the propeller in one revolution, through a solid medium. *NB.* A constant pitch angle propeller is one with blades which are flat and set at a designated angle.

Propeller Shrouds A descriptive term used to describe an encased propeller often fitted with baffle plates which are set into propeller ducting for the purpose of redirecting water flow more positively to and from the propeller blades.

Propeller Slip Considered as the difference between the actual speed of the vessel and the speed of the engine. It is always expressed as a percentage (%) and determined from the formula:

$$\text{Propeller Slip \%} = \frac{\text{Engine speed} - \text{speed of vessel}}{\text{Engine speed}} \times 100$$

Pudding Fender A round rope fender usually constructed of coir interwoven rope and packed with cork granules. They are secured to light lanyards and can be easily transported to any part of the ship to prevent damage to the ship's side shell plate, in the event of a heavy landing against a dock or quay wall.

Quarter The area off the stern up to 45° either side of the fore and aft line.

Quarterdeck A traditional term which describes that aft position from which the Master conned or controlled a sailing vessel.

Quartermaster The designated title given to that person who is steering the ship and acting as the helmsman.

Racking An athwartship's stress incurred in the ship's hull by excessive rolling action by the vessel.

Range (i) Distance off of a target;
(ii) Used to describe the laying out of moorings or anchor cables. Common in dry docks is to range the anchor cable on the floor of the dry dock, usually prior to inspection.

Range of Tide That measured value between the height of low water and high water levels.

Ranging The fore and aft movement of a vessel when moored alongside. The ship is said to be 'ranging on her moorings'. This is particularly dangerous where the ship's moorings are slack and the ship's movement could cause them to part.

Rate of Turn Describes the rate of change of the ship's course per unit time. Determined while the ship completes sea trials when new. The navigation bridge would normally have a 'Rate of Turn' indicator to permit monitoring of the ship's performance during a turning manoeuvre.

Render An old term meaning to pay out a line or the anchor cable to increase the length. An alternative term meaning the same is 'veer'.

Reserve Buoyancy The total volume of the non-submerged watertight compartments.

Resistance of the Ship's Hull The total sum of friction between the ship's wetted surface and the water, of the moving hull.

Revolutions Per Minute (RPM) The number of revolutions turned in a period of one (1) minute. In the marine environment it is generally a reference to the speed of the shaft(s) turning the propellers. The RPM being indicated on the navigation bridge by an 'RPM counter'.

Riding Cable That anchor cable which is secured to the up-tide anchor that takes the weight of the vessel when the ship is positioned in a standing or running moor.

Riding Lights An alternative name which describes the anchor lights displayed by a vessel when riding to her anchor.

Rising Tide Term used to describe when the tide is making and increasing the water depth on the flood.

Roads Generally a shortened term for 'Pilot Roads' where the vessel tends to make a landfall and attain the pilot boat station. The Roads is a focal point area for shipping and often close to narrows, where the need for the local knowledge of a marine pilot is required before proceeding.

Roadstead Similar to 'Roads' but lends to being a safe anchorage with good holding ground.

Rogue Wave A descriptive term meaning an exceptionally large wave. Recent research has shown that these are not as isolated as previously thought and in fact may occur in many geographic locations in any of the world's oceans.

Rope Guard A steel protective fitted between the hull and the propeller to prevent mooring ropes fouling in the propeller.

Rotary Vane Steering A steering system consisting of a rotor keyed to the rudder stock. Hydraulic fluid under pressure is pumped to the rotor causing the stock and subsequently the rudder to turn. The direction of the pumped fluid reflects the movement of the rudder.

Rough Sea A sea state which has considerable turbulence accompanied by wind force 5–7 on the Beaufort Scale.

Round Turn (i) A term used to describe a foul hawse, where the cables have turned about themselves with the ship passing through 720°; (ii) A term which describes the action of the vessel making a complete 360° turn. It is generally considered an extreme manoeuvre when taking action in a collision avoidance situation to evade a close quarters situation.

Rudder A vertical steering unit generally positioned at the stern of the vessel (some vessels are constructed with bow rudders where the vessel expects to conduct extensive stern first navigation). The rudder is connected to the steering systems of the navigation bridge from where it can be controlled to provide directional heading to the vessel. Some vessels would carry twin rudders, when fitted with multiple propellers.

Rudder Carrier A constructional feature fitted inboard under the tiller position, to accept the weight of the rudder stock.

Rudder Indicator An instrument on the navigation bridge that provides feedback to the helmsman showing the angle to which the rudder has moved following a helm movement. (Not to be confused with a 'Helm Indicator'.)

Running Lights The navigation lights required by law to be shown by a ship when steaming or sailing at night.

Scope The amount of anchor cable deployed, measured from the mouth of the hawse pipe to the anchor crown 'D' shackle.

Sea Anchor An improvised drogue streamed over the bow, designed to keep the vessels head to wind and reduce drift. It would only be employed as an emergency measure to prevent the unwanted movement of the vessel.

Sea Breeze A breeze which blows from the sea to the shore during the day; a land breeze being the opposite – blowing from the land towards the sea during the night time.

Sea (ships) Trials A testing and trial period for a newly constructed ship to ascertain the vessel's criteria and capabilities.

Shackle (i) A shackle length of anchor cable is defined as a length of anchor cable equal to 15 fathoms (90 feet or 27.5 metres). The number of shackles carried by vessels differs with the size of ship and trade; (ii) Shackle is a term which describes an individual fitment extensively used in anchorwork, but not excluded to just anchorwork. There are many types of shackles in operation, not all in the marine industry. Examples of shackles include: mooring shackles for securing ships to buoys; joining shackles for joining anchor cable lengths; anchor shackles for joining cable to anchor shanks.

Shallow Water Effect A form of interaction which can affect the steerage of the vessel when in shallow waters with limited underkeel clearance.

Sheer The action of turning the vessel off the line of cable when lying to a single anchor. It is achieved by placing the rudder hard over and causing the vessel to angle away, the rudder still being effective at anchor as a stream of water is passing the rudder position.

Shorten Cable A term used to describe the action of reducing the scope of the anchor cable of a vessel lying to her anchor(s).

Short Stay A description of the anchor cable of an anchored vessel, when there is a limited amount of chain cable visible above the surface, and the cable is at an acute angle to the waterline. (Long Stay describes when the cable is nearly parallel to the water line and extended.)

Sighted and Clear An expression used when heaving up the anchor to describe when the anchor breaks the surface of the water and is sighted and seen to be clear of obstructions.

Single Anchor The action of a ship going to an anchorage and deploying a single anchor. The circle of swing created with this action will be large; as opposed to a vessel mooring, which would be expected to deploy two anchors and gain a reduced swinging room.

Single-Up An order given to mooring parties to reduce the number of moorings to a manageable number (one or two) prior to a vessel; departing a berth.

Skeg The aft extension of a keel and is the deepest part of the aft structure. A sole piece of a stern frame may incorporate a skeg section.

Slack Water That interval between tides where the tidal current is very weak or non-effective, usually occurring between the reversal of the tidal flow; but it can occur at any time, about the period of the turn of the tide.

Sleeping Cable That cable which is secured to the down-tide anchor which bears no weight when deployed in a running or standing moor (see Riding Cable).

Slip Wire A bight of wire rigged to pass through the ring of a mooring buoy. It is always the last mooring out, once the vessel is secured to buoys and designed to be the last mooring released. The purpose of the slip wire is to allow the ship's personnel to control the time of departure and not be dependent on shoreside linesmen. They are rigged from each end of the vessel using a messenger and mooring boat, when the ship is secured to buoys.

Smelling the Bottom A term which describes a vessel with little underkeel clearance where the keel is close to the sea bottom. The flow of water around the hull disturbs the silt and will usually cause the water astern to be stained by the mud.

Snub Round A descriptive term for a manoeuvre, where a ship turns on its anchor when deployed at short stay.

Sole Piece The lower part of the stern frame construction which supports the bearing pintle of the rudder. When the vessel is trimmed by the stern it is that deepest part of the vessel.

Sounding That depth of water given on the nautical chart and the actual depth of water that the vessel is positioned in. An echo sounding machine or a lead line is the usual method of obtaining the water depth. (The term is also used to gauge the depth of fluid in a tank.)

Spoil Ground This is a dumping area, usually marked on the navigation chart and an area that should be avoided especially for anchoring.

Spring Tide A tide with maximum range as a result of the combined effect of the sun and moon's position. It occurs twice per lunar month.

Spring Wire A steel wire mooring line employed in opposition to head lines and stern lines to prevent the vessel ranging when alongside the quay.

Squat A form of interaction often experienced in shallow water areas like rivers and canals, where the vessel is observed to experience bodily sinkage and sit lower in the water than would normally happen as in deep water. A vessel may squat by the head or by the stern but it is a more common occurrence to squat by the stern. Squat is directly related to the speed^2 of the vessel.

Stand On Vessel That vessel which is required by the COLREGS to maintain her course and speed when given the right of way by the regulations.

Starboard Defined by the right of the ship when facing forward (opposite to the port side of the vessel). Also used as a term when giving helm orders when manoeuvring the ship. The US uses left or right rudder to express a desire to move to Port or Starboard, respectively.

Steerageway A term which describes that the vessel is still responding to the helm when the vessel is at minimum speed.

Stem Anchor An anchor set into a position on the stem of the vessel. This is not a common arrangement compared with ships which are usually fitted with two bow anchors.

Sternway An expression that describes a vessel moving astern under her own power or with her own machinery stopped.

Stockless Anchor A patent anchor common to every-day use which is stowed inside the hawse pipe of an ocean-going vessel. There are many variations of modern designs currently widely used in the marine environment which do not carry the old fashioned cross bar stock.

Stopper A length of rope or chain employed to temporarily take the weight of a rope or wire, while it is transferred from a winch to secure cleats or bollards.

Stopping Distance Defined as the minimum distance that a vessel may be seen to come to rest over the ground. The distance is usually determined from a ship's trials when the vessel is new. Test runs will normally provide the stopping distance: (a) from full ahead after ordering the main engines to stop; (b) from crash full astern (emergency stop).

Storm Moorings Shore side moorings which are secured to the vessel in the event of anticipated bad weather while the vessel is alongside. More common to Ports of the Pacific Rim, which experience heavy swell action.

Storm Surge An increase in the level of water along the coastline due to strong onshore storm winds. Negative storm surges can also be experienced some time after the passing of the storm, producing less tidal heights than predicted.

Stranding When a vessel has grounded for a period of time it is said to be stranded for the purpose of Marine Insurance.

Stream Anchor A light anchor sometimes carried at the stern of the vessel. Alternatively called a stern anchor or kedge anchor.

Surge A term used to describe a mooring rope being allowed to slip about a turning winch barrel. Synthetic ropes should not be surged because the generated heat could destroy the fibres of the rope.

Swinging Room The circle area scribed by a vessel when lying at anchor that the vessel will turn through from one tide to another.

Swivel Piece An anchor cable fitment which may be incorporated in the ganger length of the anchor cable to prevent kinks forming in the cable. Alternatively, it may be the term used to describe a 'Mooring Swivel Piece' which is set into the anchor cable when a vessel moors to buoys to prevent anchor cables becoming fouled. It would normally be employed if the vessel was being moored for a lengthy period of time.

Synchronizing A term used to describe the movement of the vessel when rolling or pitching, when the ship's movement matches the period of encounter of a wave.

Synchro-Lift A system of dry docking ships which employs an elevating platform in the single dock. Once the vessel is lifted by the elevator it is pushed and/or towed into a docking bay. The system allows several ships to be docked at the same time and does not prevent other vessels using the elevator docking operation. With the ship on, the lifting platform is raised by mechanical means (winches on dock sides) and limits the size of vessel that can use the facilities.

Synoptic Chart A weather chart showing weather patterns, fronts and pressure systems.

Tactical Diameter That greatest diameter scribed by the vessel when commencing and completing a turning circle.

Thrust Block An engine room fitting that receives the thrust from the propeller. It incorporates the thrust bearings.

Thruster A powered propeller or jet, positioned either forward or aft in the ship. Its purpose is to aid the turning motion of the vessel when manoeuvring.

Tidal Range The average difference between the high and low water, assessed over a period of a month or more.

Tide Rode An expression which describes a vessel at anchor lying in the direction of the tidal flow as opposed to 'Wind Rode' where the vessel is lying to the direction of the wind.

Topmark An additional shape carried by a buoy to emphasize the type and function of the buoy.

Towing Horse An athwartship's aft arrangement which is designed to act as a moveable lead, in the stern region of the towing vessel.

Towing Light A yellow navigation light carried by a tug when engaged in towing. The light is carried at or near the stern and has the same characteristics as the normal stern light.

Tractor Tug A tug fitted with an omi-directional propulsion system, e.g. Voith Schneider, cycloid thruster. Usually operates as a highly manoeuvrable harbour tug.

Transfer Defined by that distance gained by a vessel engaged in a turning manoeuvre which is perpendicular to the original course.

Transverse Thrust An expression that describes the imbalance from the water flow about a propeller causing a vessel to pay off to one side or another. Most pronounced when operating astern propulsion.

Trim The difference between the forward draught and the after draught. Ships generally trim by the stern to provide ease of steering.

Trough The lower dip between wave crests is termed the trough of a wave.

Tsunami A Japanese word, often incorrectly referred to as a tidal wave. A wave surge usually generated from an under surface disturbance like a sub-sea earthquake, causing major damage when reaching the shoreline.

Tunnel Thruster A type of 'Bow Thrust Unit' which passes from either side of the ship to provide thrust to port or starboard. May also be employed as a stern thruster.

Turn Short Round A ship's manoeuvre which endeavours to turn the vessel in its own length.

Typhoon A tropical storm common to the Western Pacific Ocean, derived from the Chinese word Tai-fung.

Underfoot A term used to describe an anchor being released just under the stem or the forefoot. Generally used to gain reduced movement of the ship's head when at anchor.

Underkeel Clearance A measurement of the amount of water under the ship's keel. The value is obtained from the echo sounder with corrections applied.

Underway Defined by the Regulations for the Prevention of Collision at Sea and refers to a vessel not at anchor, made fast to the shore or aground.

Up and Down A term used to describe the direction of the anchor cable being at right angles to the water surface.

Variable Pitch Propeller A propeller with blades where the angle of pitch can be altered. Also known as Controllable Pitch Propeller (CPP).

Variation That angle between the bearing of the True North Pole and the magnetic North Pole. The angle will vary with the ship's position on the earth's surface and can be found from the nautical chart. It is also coupled with deviation to provide the value of the Compass Error.

Veer A term used to describe the paying out or slacking down of a line or anchor cable. To veer anchor cable meaning to pay out and slacken the cable.

Vessel Traffic System (VTS) A system that controls shipping in and around coastlines and congested waters. It is usually operated by coastguard organizations or other respected authorities.

Voith-Schneider Propellers A propeller action fitted to a vertical shaft. The system has a number of vertical hanging blades caused to rotate in a horizontal plane generating vessel directional movement.

Wake The disturbed track of surface water left by the ship's propeller(s) as she moves ahead.

Wake Current A forward movement of water caused by hull friction from the propeller region, when the vessel is moving ahead. It is of small significance but does adversely affect the efficiency of the propeller.

Walk Back An expression used to describe the paying out under control of a mooring line or anchor cable.

Warp An alternative term to describe a ship's mooring line.

Warping The action of moving the ship by means of the ship's mooring lines. (Engines not usually being employed to move the vessel.)

Wash Turbulent water as caused, say, by a rotating propeller.

Watch Shipboard duties are contained within a shift or watch system. Navigation, engine room and anchor duties are all carried out through structured periods of time known as watches.

Water Jet A modern method of propulsion or thruster unit currently being fitted to high speed craft.

Wave Height That vertical distance between the crest of a wave and the lower part of the trough.

Wave Length Is defined by the distance between two adjacent crests of waves.

Way When a vessel starts her main engines, when fitted with a conventional fixed pitch propeller and commences to move forward, she is said to be '*gathering way*'. The term '*making way*' defines when the vessel is moving through the water, when under her own power. '*Steerage way*' is an expression which describes when the speed of the vessel will still effect and obtain a correct rudder response, causing the desired movement of the ship's head. '*Sternway*' is when the vessel is moving over the ground in an astern direction.

Weather Side That side which faces the wind (when referred to a ship).

Weather Deck The uppermost, uncovered deck of a ship, which is exposed to the weather.

Weigh A descriptive term to express the lifting and raising of the ship's anchor.

Wide Berth A term to describe giving a navigation hazard adequate clearance.

Windlass The name given to a heavy duty mooring winch in the fore part of the vessel engaged as an anchor cable lifter. They are generally multi-purpose, providing warping barrels for mooring rope use.

Wind Rode A vessel is described as wind rode when she is riding to her anchor head to wind.

Windward That side on which the wind blows and faces the prevailing weather.

Yaw A term used to describe the movement of the ship's head away from her designated course. The movement can be to either port or starboard and is influenced by a following wind, or sea conditions. It should be noted that a vessel may 'yaw about' when weather conditions are from another direction other than from astern. The vessel may even 'yaw about' the anchor position when moored to a single anchor. The movement should not be confused with 'Sheering'.

Tidal reference

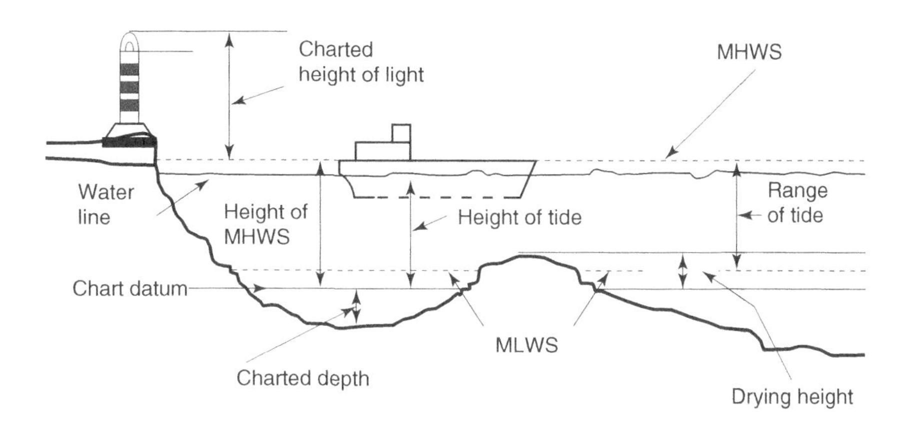

MHWS = Mean high water springs MLWS = Mean low water springs

Tidal terms and reference to chart datum

Introduction

This work has been produced to meet the needs of the serving Merchant Navy Officer when at sea and the marine student when ashore, studying for marine qualifications. The book combines the changes in hardware and the handling skills required to manoeuvre today's modern shipping safely, within the developing maritime environment.

Each chapter covers a specific topic area, including routine and emergency manoeuvres, which allows the reader to visualize activities associated with all aspects of Ship Handling. The topic is an aspect of Seamanship, and some prior knowledge by the reader has been anticipated.

The volume is expected to stand alongside its sister work *The Seamanship Examiner*. It is not meant to be a substitute for the real time experience gained on board the deck of a vessel at sea. Nor can it hope to introduce the reality effects of wind and current experienced by the practising mariner. However, it may possibly provide the theoretical knowledge to support practical skills while at the same time giving confidence to the examination candidate.

The work will not remove the hazards associated with shipping, but may reiterate the need for continuous awareness amongst our senior and the up-and-coming junior ranks, who can expect to drive our ships of the future. I have long felt that learning should be fun and an enjoyable exercise. If the topic is boring, interest is lost and nobody gains. It is hoped that the use of this work, will be a beneficial reference to the practising mariner and keep the seas clean and safe for all.

Good Sailing!

1 Ship handling and manoeuvring

Introduction; Manoeuvring and handling scenarios; Turning short round; Snub round; Berthing and unberthing; Entering a dock; Use of mooring lines and deck equipment.

Introduction

It is impossible for any text, or other simulation to imagine that it could substitute for the practicalities of real time ship handling operations. Nothing can be a substitute for the real thing. However, the theory behind ship manoeuvres can be explained but it is up to the practitioner to then take full account of the wind and tidal effects in a real-life situation.

Ship handling theory is a vast topic in its own right because not only are there numerous manoeuvres but so many variants within those manoeuvres (such as those effected by single right hand fixed propellers, twin screw vessels, ships with controllable pitch propellers, ships with tugs and without tugs, good weather or bad weather conditions prevailing, with tide or without tide, etc.).

The practitioner can take heart from the fact that the more handling and the more manoeuvres that are attempted, the greater will be the expertise that is to be gained. It is hoped that this chapter will deal with the fundamentals of ship handling and provide theoretical principles of operation covering most of the more common situations.

Where modern hardware (like bow thruster/stern thrusters or controllable pitch propellers) are used, alternative manoeuvres are easily employed; although it is appreciated that some vessels are fitted with only basic manoeuvring aids.

Ship handling has always been placed firmly in the hands of the ship's Master; which is, without doubt, unfortunate in many aspects for the future. Especially so when the industry expects that the newly promoted Master should become an expert ship handler, virtually overnight, often with no previous experience. A distinct lack of opportunity and positive training in the subject has long been recognized as a failing point of the maritime sector. It would certainly be helpful and advance education, if Masters were to encourage their junior officers to gain hands-on experience, whenever safety and time allows.

Aspects of ship handling

The men who handle our vessels are not born expert ship handlers, neither are they made from a mould. Usually, they are self-taught and become well-practiced over time. They have a variety of elements under their control, such as: engines and relevant speed control, helm and steering gear effecting rudder(s), bow and stern thrust units if fitted, stabilizers, anchors and moorings. Also 'tugs', assuming they respond to the directions of the conn. To some extent, draught and trim of the vessel can be controlled within limits, provided that the vessel is undamaged.

What of course makes the task of the ship handler so challenging, is that many elements are not under his or her control but still have to be catered for. Clear examples of this are the weather, tide heights and times, depth of water and respective underkeel clearance, manmade objects like bridges, geographic obstructions as with narrows, etc. The person handling the vessel in confined waters will employ all elements under their control as well as the elements that lie outside their control, e.g. the wind. To achieve the objective any aspect including 'luck' is usually gratefully accepted.

The forward mooring deck of a Class 1, passenger vessel. The deck is fitted with a centre pipe lead, with triple roller fairleads either side. International roller fairleads, with Panama leads are sited to port and starboard, set into the bulkheads. The wide beam ship carries split windlasses, each with tension winch and warping drum incorporated. 'Old men' roller leads are also seen amongst the sets of bitts on the deck forward of a spare anchor fixed to the deck on the centre line.

Turning short round (right hand fixed propeller)

Alter the ship's head to move to the port side of the channel, as this would gain the greatest advantage when operating astern from transverse thrust, during the turn.

1. Dead slow ahead on engines and order the helm hard to starboard.
2. Stop engines, wheel midships.

3. Vessel still moving ahead making headreach.
   Full astern, wheel amidships, until the vessel gathers sternway, then stop engines. The effect of transverse thrust would generate a tendency for the bow to move to starboard and the stern to move to port (with the ships bow in the centre of the channel, where the flow is the strongest the tide effect would tend to push the bow round to starboard).

4 & 5. Wheel hard to starboard, engines full ahead to achieve the reverse heading.

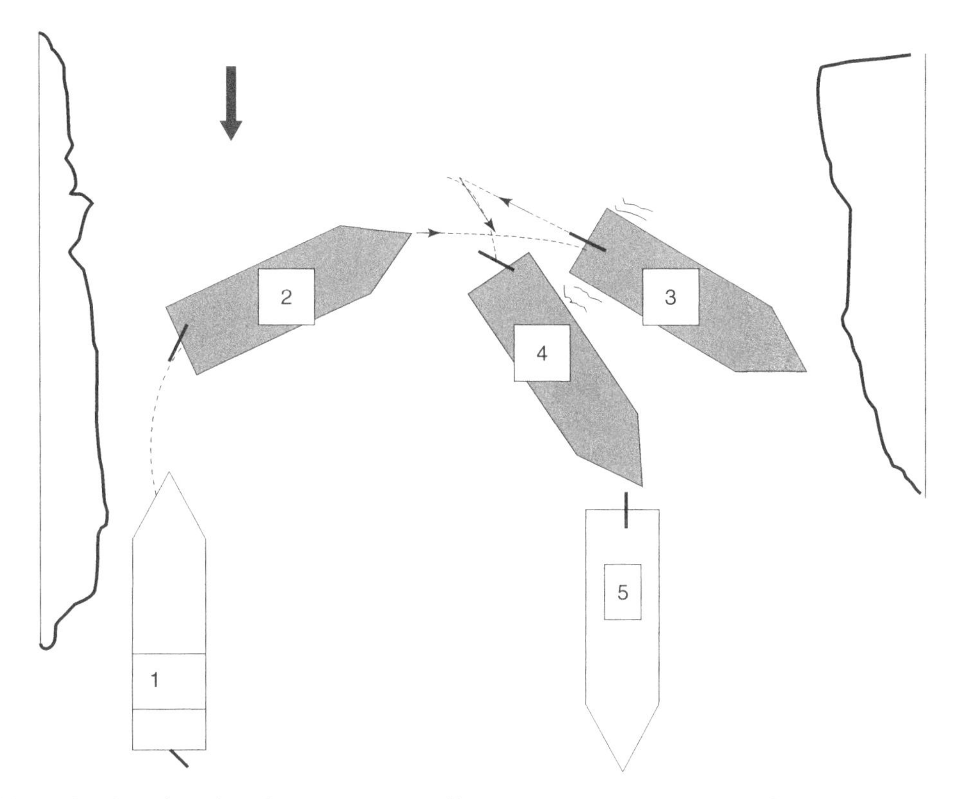

The objective of turning short round is to effect a tight turn within the ship's own length or as near as possible to within its own length.

Snubbing round (tide astern)

The objective of this manoeuvre is to turn the vessel where restricted sea room exists.

The turn employs the use of a single anchor and can be made turning to port or starboard. It is employed to turn the vessel to stem the tidal stream or can be used when berthing, leaving the anchor deployed to heave the vessel off the berth when clearing the berth.

1. Position the vessel on the port side of the channel, with the tide astern. Have the starboard anchor walked back, ready for deployment at short stay.
2. Helm should be placed hard to starboard, engines on stop. Let go the starboard anchor at position '2', to a short stay.

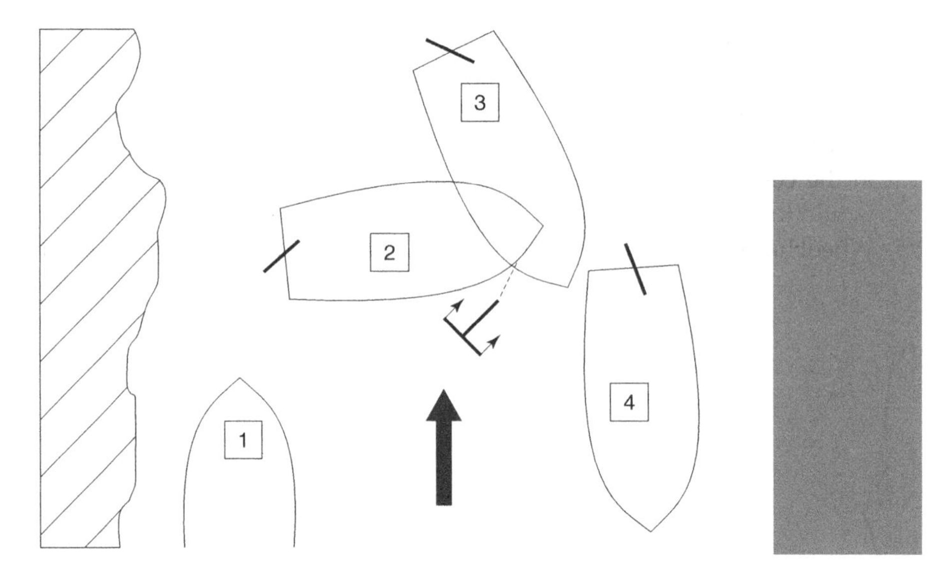

Turning (snubbing round) to starboard.

3. Check the anchor cable and keep the anchor at short stay. The momentum on the vessel should carry the stern through 180° with the bow being held by the cable.
4. The helm should be placed hard to starboard and engines on half ahead to overcome the tidal effect. Engage the windlass gear and recover the anchor, having turned the ship's head into stemming the tide.

When using the manoeuvre to turn off the berth and go alongside, the anchor cable would be paid out more to allow the vessel to close the berth. Once alongside, the cable would be walked back to the 'up and down' position so as not to obstruct the channel.

NB. The manoeuvre can also be employed by passing the bow or stern through the wind. The tide effect being the main force pushing the stern around.

Berthing and unberthing

Berthing port side to the quay – right hand fixed propeller – calm conditions

1. Approach the berth at an angle of about 25°, engines dead slow ahead.
2. Stop engines on the approach taking account of the headway that the vessel will carry.
3. Engines astern. Transverse thrust would cause the stern to swing to port and the ship would gradually stop parallel to the berth.
4. Stop engines. Send away head and stern lines and make fast.

NB. If the angle of approach is larger than suggested it may be necessary to use a small amount of starboard helm in position '3' in order to start the stern swing in towards the quay. Excessive helm use would generate a too fast stern swing.

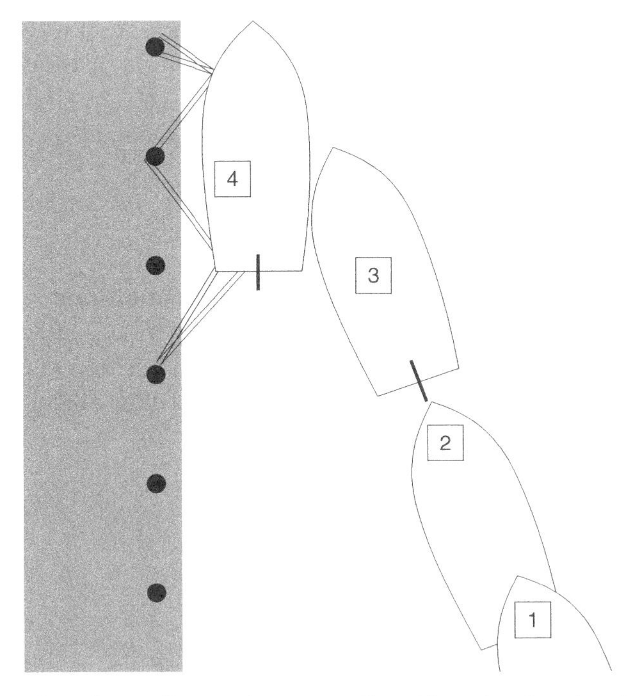

Astern movement of engines will cause transverse thrust to swing the stern towards the quay.

Berthing starboard side to quay – right hand fixed propeller – calm conditions

1. Approach the quay at a shallow angle, say at about 15°, engines dead slow ahead. Stop engines on approach taking account of the headway which the ship will carry.

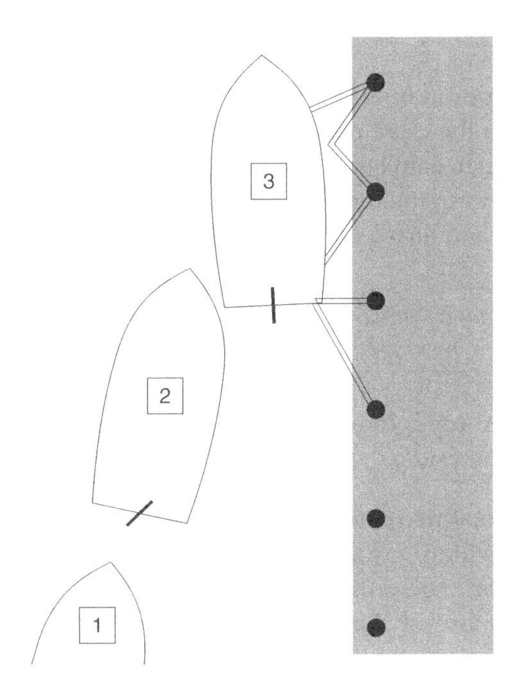

2. Approaching the berth, apply port helm to cause the stern to swing towards the berth. Engines astern to stop the ship and the effects of transverse thrust will check the stern swing.
3. Stop engines. Send away head and stern lines and make fast.

NB*. In the event that there is limited room ahead of the vessel, the forward spring line should be sent first.*

Berthing into strong offshore wind – slack water conditions

1. Approach the berth at a steep angle to reduce the windage effect on the vessel.
2. Prepare a stern line to be passed from the forward position (assuming no mooring boat is available). Approach the berth at a dead slow ahead speed.

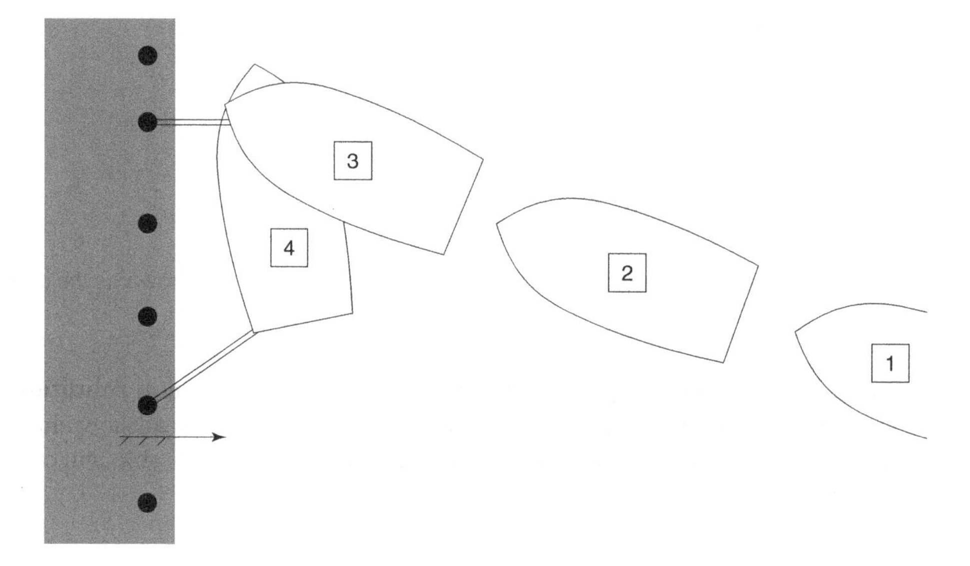

3. Stop engines, on approach, then engines astern to stop the bow just off the berth. Pass a head line to the quay using a heaving line from the wharf, rather than a ship's heaving line, to gain benefit of wind.
4. Pass the stern line from the forward position and carry the mooring up the quay. Ease the head line and heave on the stern line to bring the vessel alongside.

> **Comment**: A mooring boat employed to carry the stern line ashore would eliminate the need to pass the stern mooring forward.

Once alongside, breast ropes fore and aft would reduce the possibility of the vessel being blown off the quayside.

Berthing port side to, with a strong onshore wind

1. Stem the tide at position '1' rudder hard to starboard and engines half ahead.
2. Attain a position off the berth and parallel to the berth, with the port side well fendered (possible use of the offshore starboard anchor may be desirable for departing the berth, with the same direction of wind).

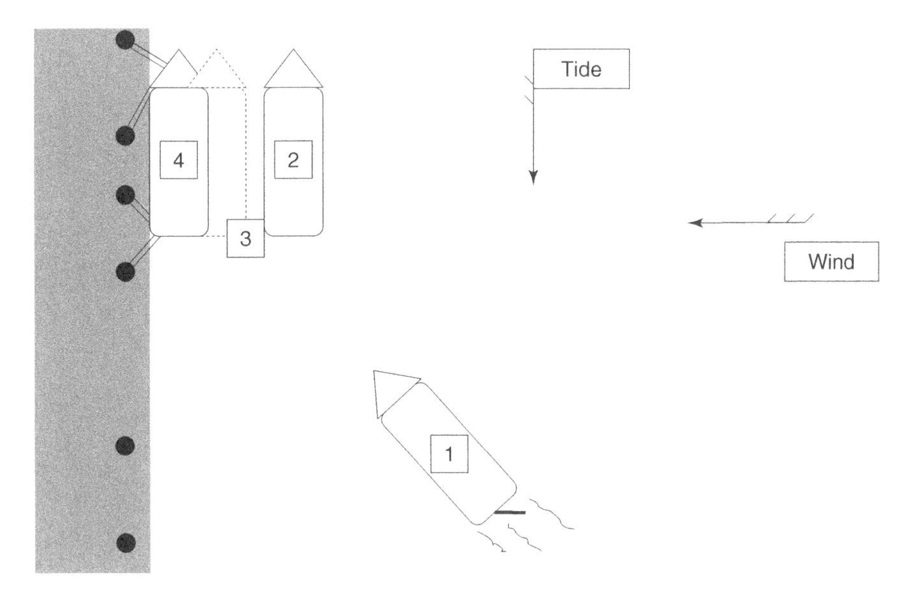

Berthing for high-sided vessels.

3. High freeboard vessels will benefit from the wind on the beam and allow the vessel to close the berth at position '3'. Run head lines and stern lines fore and aft.
4. As the vessel lands alongside the quay, pass and secure fore and aft springs and adjust the ships position to suit with head and stern lines. Once secure, if the offshore anchor has been deployed, walk back the cable to an up and down position.

Example vessels: Container ships, Ro–Ro's, Passenger Liners.

NB. *The use of the offshore anchor can clearly check the rate of approach of the bow, but the use of engines and rudder against the angled direction of the anchor cable may be needed to keep the stern parallel to the quayside and ease the landing.*

Leaving the cable in the up and down position is to avoid the chain obstructing the channel for passing traffic, while at the same time providing a useful means of heaving the ship off the berth against the onshore wind, when departing.

Berthing port side to (for vessels with windage area aft) – strong onshore wind

1. Approach the berth at about a 60° angle. Stop the vessel off the berth with the bow level with the centre of the berthing position. Let go the offshore anchor at short stay. To control the stern against the wind, use rudder to port and engines ahead. Dredge the anchor towards the berth.
2. As the vessel approaches the berth, pay out the anchor cable.
3. When the bow is just off the berth, hold on to the anchor. The vessel will pivot at the hawse pipe and the stern will swing rapidly towards the quay.
4. As the stern is approaching the quay, engines ahead to check the stern swing. Stop engines and run lines ashore fore and aft.

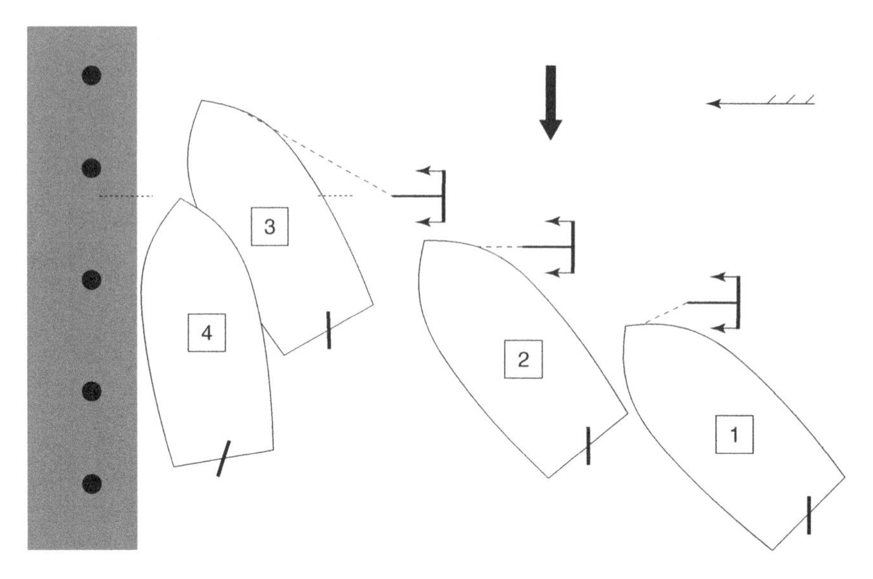

Berthing using an offshore anchor.

Berthing using an offshore anchor
For a variety of reasons, many ships will employ an offshore anchor when berthing.

The vessel 'Pedernales' lies starboard side to, secured by two head lines and a rope spring at the forward end. The offshore anchor has been deployed during the berthing operation and the cable is seen walked back to an up and down position so as not to cause obstruction. Panama leads are seen centre an either side together with triple roller fairleads accommodating the lead for the head lines. Mooring ropes are soft eye, multi-plait polypropylene hawsers.

Some vessels will use the anchor to permit turning off the berth, so as to be heading in the desired direction ready for departure. Other vessels may have to berth with a specific side to, for cargo requirements. While others, realizing the direction of an onshore wind, will use the offshore anchor with a view to heave the vessel off the quay when departing.

In virtually every case it would be considered poor seamanship to leave the anchor cable stretched outwards, after the vessel has berthed. The harbour authority would generally not permit such an obstruction as it would tend to impede the movement of other traffic.

To this end it is normal practice that the cable would be walked back to the up and down position once the vessel is secured alongside.

Berthing starboard side to – tide ahead – right hand fixed propeller

1. Stem the tide and approach the berth using engines ahead to maintain position.
2. Apply a little starboard helm to cause the bow to cant towards the berth. Then steady the ships head. The vessel could expect to move bodily towards the berth.

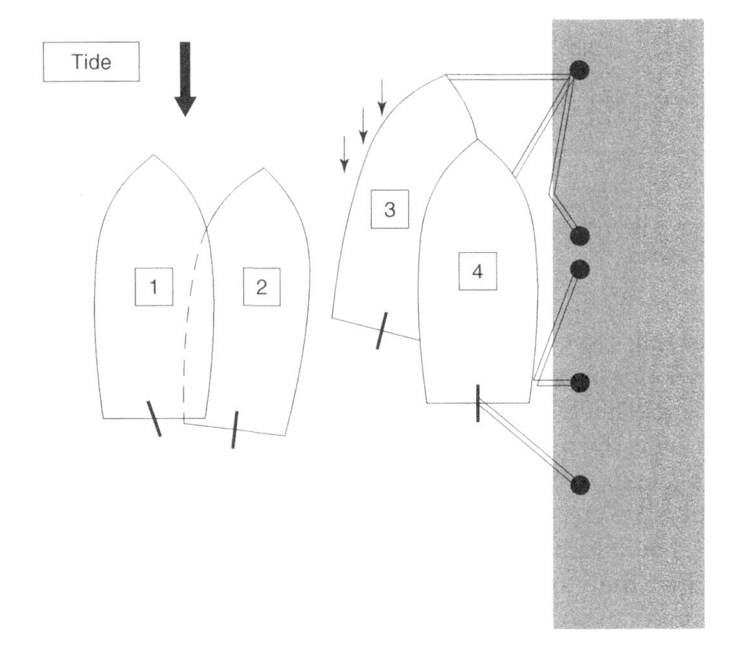

3. Just off the berth, bring the vessel head to tide and send away a head line with an aft spring.
4. Once alongside, stop engines and make fast with head lines, stern lines and springs.

> **Comment:** If an 'Offshore Wind' is present, the use of engines with the port helm may be necessary to cause the stern to close the quay to pass stern lines. Alternatively, a mooring boat could be employed.
>
> When an 'Onshore Wind' is present it may be necessary to ease the headline once landed, to allow the stern to close from the effects of wind and tide.

Ebb/Flood swing for berthing

Assume your own vessel is in a position '1' stemming the ebb tidal stream. A mooring boat is available, and the vessel is fitted with a right hand fixed bladed propeller.

1. The vessel must be manoeuvred to stem the tidal current as in position '1'.
2. Manoeuvre the ship to a position '2' parallel to the moored vessel 'A'.

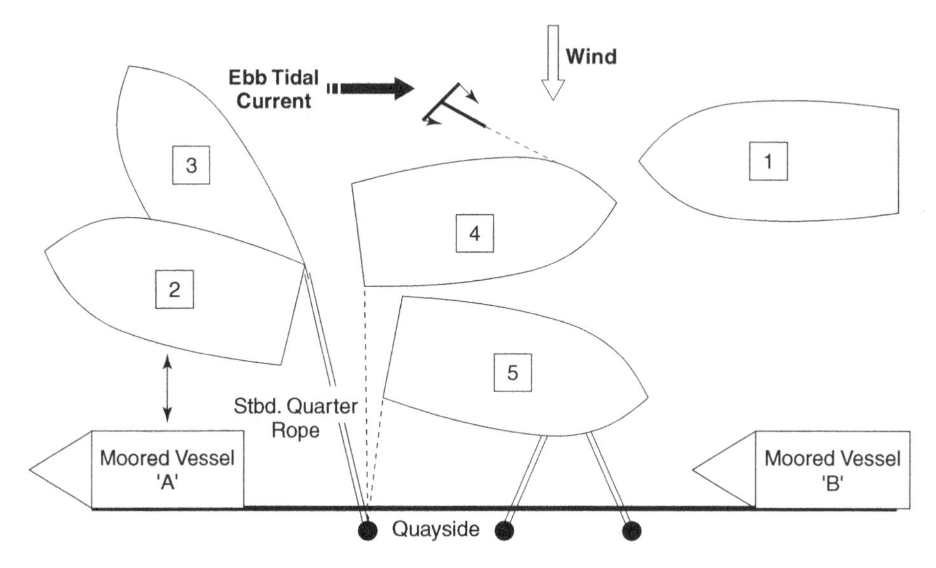

Berthing your own vessel, starboard side to, between two ships already secured alongside, with an onshore wind.

3. Run the best ship's mooring rope from the starboard quarter to the quayside with the aid of the mooring boat '2' and keep this quarter rope tight, above the water surface.
4. Slow astern on Main Engines – then Stop. This movement should bring the vessel's stern in towards the quayside by means of 'Transverse Thrust'. The bow should therefore move to starboard, outward to bring the current onto the ship's port bow. Position '3'.
5. The vessel should turn with current effective on the port side and no slack given on the quarter rope to complete an 'Ebb Swing' Position '4'. The wind should expect to affect the port bow, blowing the ship rapidly towards the quayside. To check this movement towards the quay, let go the offshore anchor.
6. Run the forward spring line to check ahead movement (position '5'). Run the head line and draw the vessel alongside from the fore and aft mooring positions, easing the weight on the anchor cable as the vessel closes the quay.

NB. *Where the manoeuvre is required when an 'onshore wind' is present, use of the offshore anchor would expect to reduce the rate of approach towards the berth (as shown).*

If an 'offshore wind' is present the first bow line would probably need to be carried aft to allow it to be passed ashore or alternatively use the same mooring boat which was initially employed. The effects of the tide would also cause the forward end to probably close towards the quay.

Unberthing – starboard side to, with offshore wind, no tide

1. Single up to head line and stern line (or breast lines).
2. Ease the head and stern lines to allow the vessel to be blown off the quay. When the stern is clear of the quay, hold on to the aft line and allow the bow to come off the quay a little more.
3. Once clear of the quayside, let go bow and stern lines and engage engines and helm.

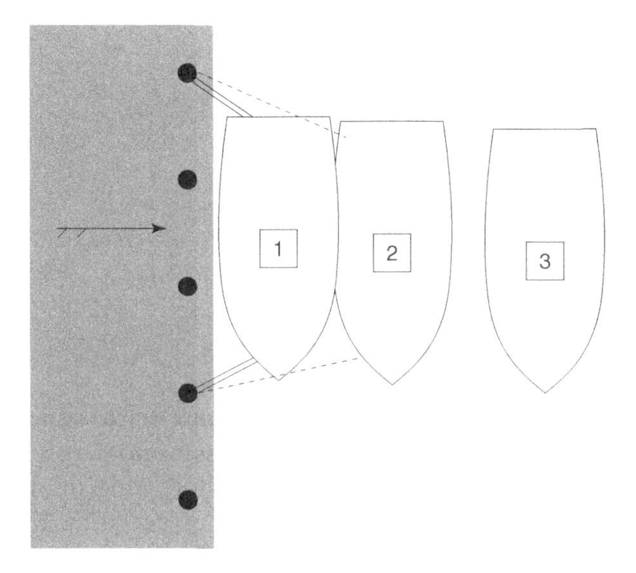

> *Comment*: This method is recommended for high sided vessels like car carriers and Ro–Ro's, which have their superstructure exposed over and above the quay height.
>
> Deep laden vessels with low superstructure may have to use a double spring mooring forward and spring the stern of the vessel off the quay into deeper water.

Unberthing – port side to, tide ahead, no wind

The objective is to clear the berth when a tidal stream is ahead of the vessel. The action allows a wedge of water to flow between the dock wall and the ship's side so forcing the vessel off the berth.

1. The vessel should be singled up to a head line, and an aft spring.
2. The aft spring line should initially be kept tight, while the head line is slacked down. The tidal stream effect would pivot the vessel about the spring and cause the bow to move off the berth. The weight of stream water moving between the berth and the ship's side, forces the stern a little away from the dock.
3. Dead slow ahead on engines and let go forward. Stop engines and let go aft.
4. Engines ahead to clear the berth into the tidal stream.

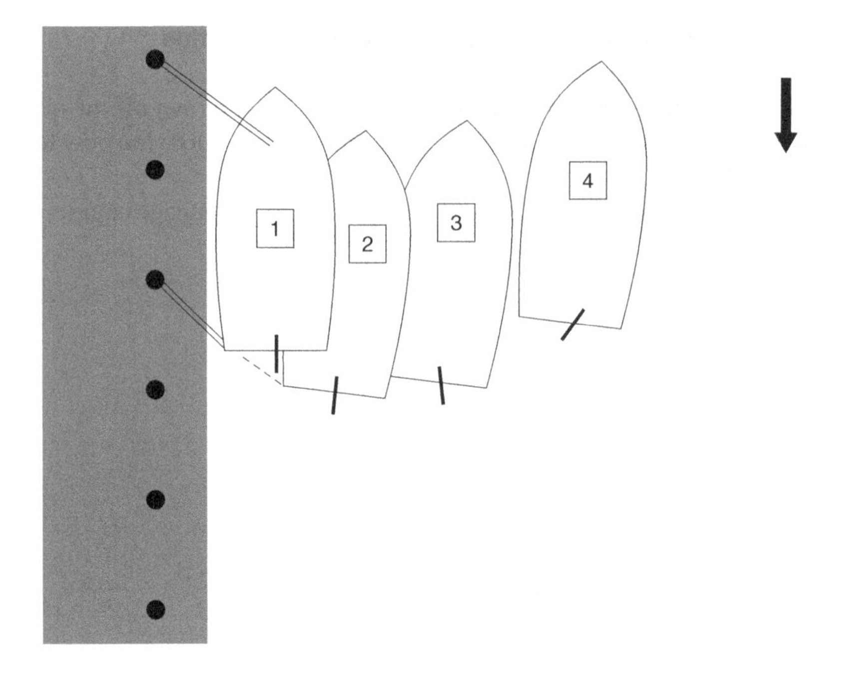

Unberthing – starboard side to, no wind and slack water conditions

The objective is to clear the berth and take the vessel into deep water. Where slack water conditions prevail, an alternative method for using the tidal/stream flow must be employed to manoeuvre the vessel clear of the berth. The prudent use of mooring lines can achieve initial movement of the vessel, so that the propeller use can be utilized.

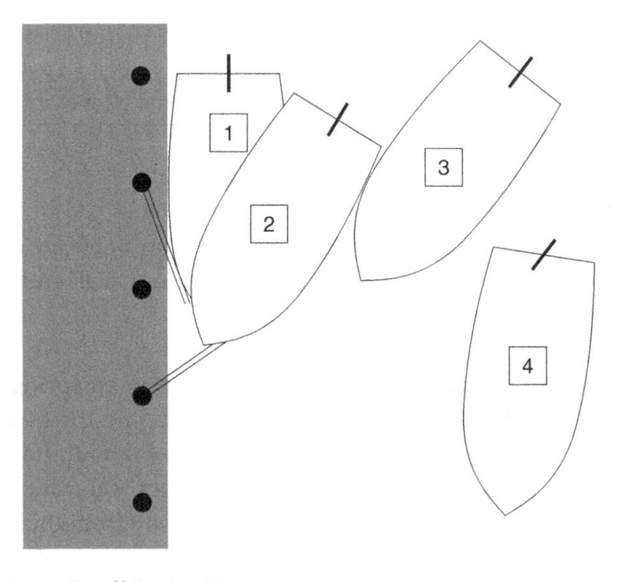

Use of moorings to angle off the berth.

Assuming a right hand fixed propeller

1. Single up to a forward spring and an offshore headline. Let all lines go aft.
2. Heave on the offshore headline to tension the spring and place engines at dead slow ahead. The stern would be expected to turn outwards away from the berth.
3. Once the stern is angled away from the berth, place rudder amidships and let go the head line. Operate astern propulsion and as the vessel comes astern the spring goes slack and can be let go.
4. As all lines are cleared, the vessel increases astern propulsion with rudder amidships. The effect of transverse thrust will cause the stern to move to port.

Unberthing – port side to, no wind and slack water conditions

The objective is to clear the berth and take the vessel into clear water where the initial effects of transverse thrust would compromise the use of the right hand fixed propeller.

1. The vessel should be singled up to an offshore head line and the forward spring (the spring could be doubled up for this manoeuvre).

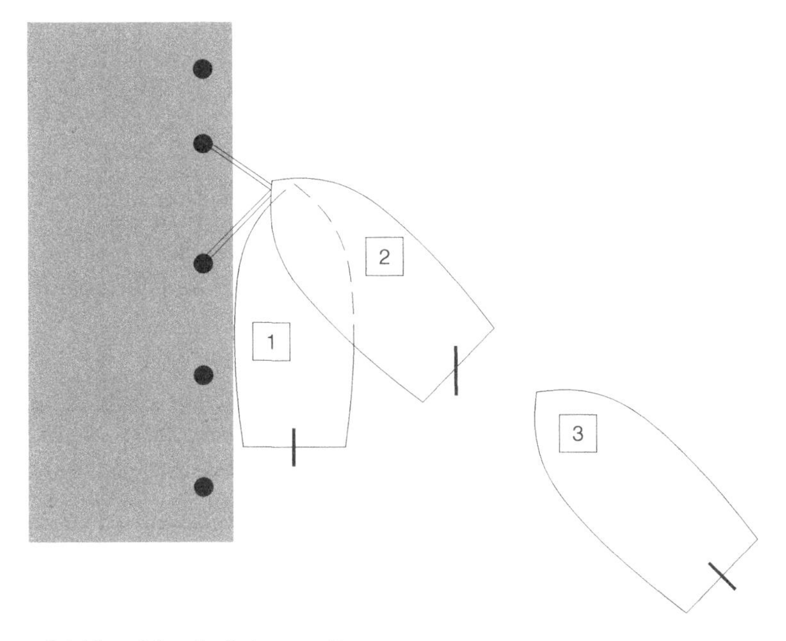

Assuming a right hand fixed pitch propeller.

2. Heave on the offshore head line to tension the spring and go dead slow ahead on engines. This action should cause the stern to move outward to starboard clear of the berth.
3. Once the stern is angled away from the berth, let go the head line and forward spring. As the lines are cleared, place the rudder amidships and the engines half or full astern. Such action would cause the transverse thrust effect to turn the vessel parallel to the berth. This would place the propeller in deep water and permit the unobstructed manoeuvring of the vessel.

Unberthing – tide astern – starboard side to

1. Single up to a forward spring and a stern line.
2. Ease stern line to tension the spring. The vessel will pivot on the spring and the stern will come off the quay.

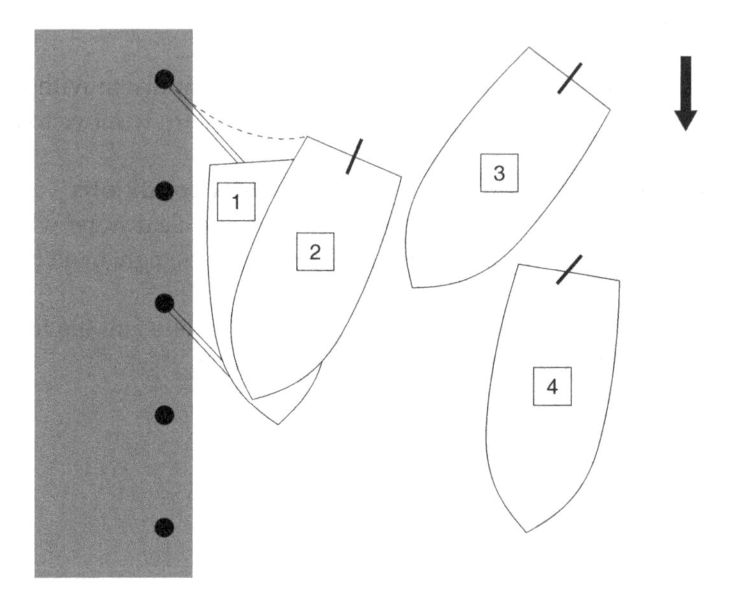

3. Once the stern is clear of the quay. Engines astern and let go forward. Stop engines and let go aft.
4. Rudder amidships, full astern into deeper water.

Comment: The transverse thrust of the engines when going astern would cause the stern of the vessel to come further off the quay into deeper water, prior to proceeding up or down river.

Unberthing starboard side to a 'T' jetty with onshore wind, no tide

1. Single up to an offshore headline and double the forward spring. Let all lines go aft.
2. Heave on the head line to tension the spring. Rudder hard to starboard, dead slow ahead on engines. Heave on the offshore headline at the same time to force the stern into the wind direction.
3. Once the stern is off the quay, place rudder amidships, let go headline and spring smartly and operate engines at full astern, towards position '3' (use of engines at full astern should cause the bow to clear the quayside quickly).

Comment: This manoeuvre is easier with a tide ahead or astern, where the wedge of water setting between the quayside and the ship will have a more pronounced effect than the wind and force the vessel away from the Jetty.

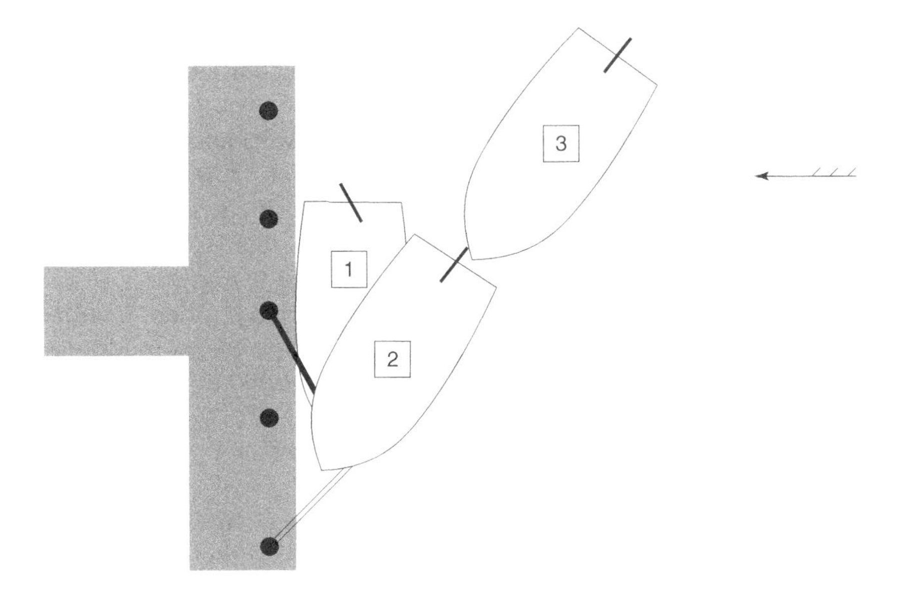

Entering a dock

Entering a dock – from a tidal estuary (no tugs available)

The objective of the manoeuvre is to enter the dock safely from a tidal estuary. It would be first expected that the vessel would stem the tidal flow direction, prior to attempting to enter the dock area.

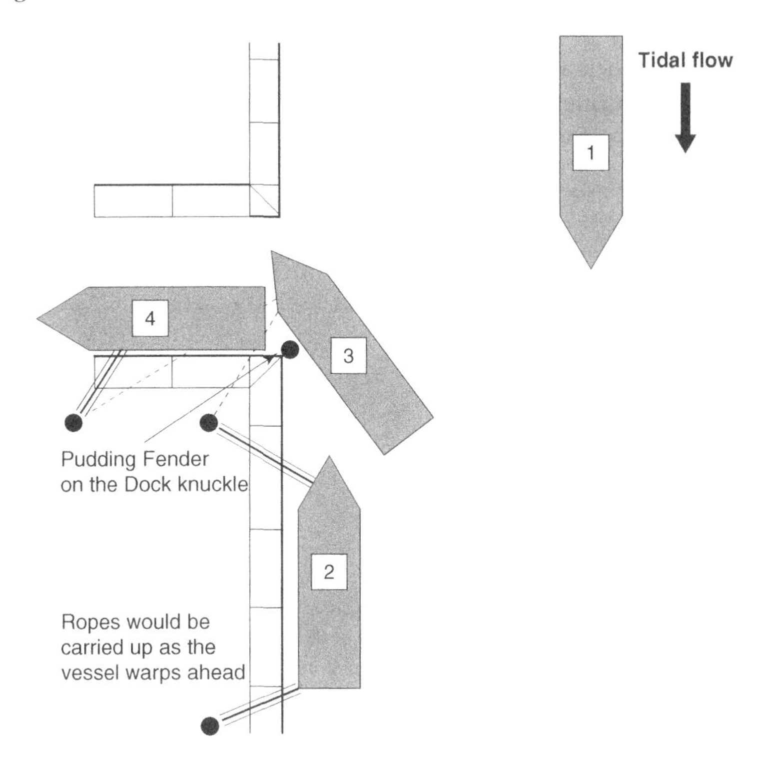

Turning the vessel into a dock is achieved by first stemming the tide usually by turning short round or snubbing round on the ship's anchor, assuming that the vessel initially has the tide astern.

The position of the vessel '1' is seen as in the exposed water area with the tide astern.

Once the vessel has turned to stem the tide, position '2' is achieved by laying alongside the berth below the dock entrance. This position is held with mooring lines to prevent the vessel ranging on the quay. A breast line aft, with two head lines would be normal practice.

The objective is achieved by warping the vessel ahead to position '3' with or without the use of engines, to turn the vessel about the knuckle. A 'pudding fender' being employed on the knuckle and mooring ropes 'carried up' to cause the vessel to enter the dock to position '4'.

Procedure to enter docks

1. Stem the tide and manoeuvre the vessel to a position alongside the berth.
2. Pass moorings fore and aft, with the intention of using these moorings to warp the vessel ahead into the dock entrance.
3. A 'pudding fender' should be readily available for use between the ship's side and the knuckle entrance of the dock, as the vessel is warped ahead and around the knuckle.

NB. *Dead slow ahead on engines could be used to cause the vessel to move towards the dock entrance. Failing this, the use of moving the vessel by the warps is usually adequate; the moorings being 'carried up' with the forward movement of the vessel, towards the dock entrance.*

4. Once inside the dock area and the ships aft part is clear, the dock gate (Caisson) can be closed. Moorings fore and aft would be retained to prevent the vessel from ranging against the dock wall.

Comment: Where the dock is narrow, the opposite side of the dock entrance can also be used to accommodate moorings from either bow. However, such use would require additional shoreside linesmen.

Entering a dock (with tug assistance)

Depending on the number of tugs available, the deployment and use would be determined by the needs of the vessel and its navigational state, i.e. dead ship without power, would need more than one (1) tug to be engaged.

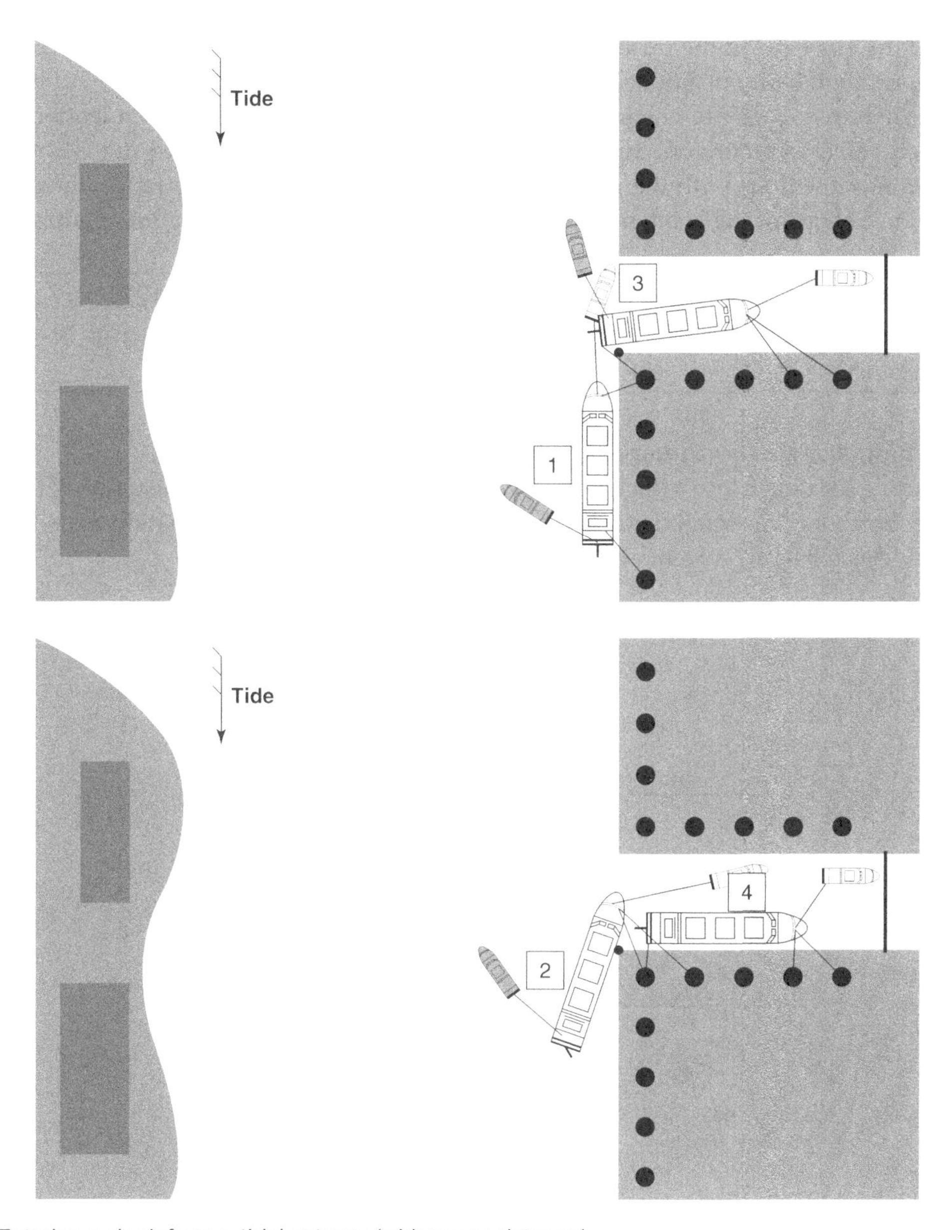

Entering a dock from a tidal estuary (with tug assistance).

Where only one tug is employed the main consideration must be the critical period as the parent vessel is turned about the knuckle to enter the dock. The direction of tidal flow will clearly push the vessel hard to the concrete knuckle. Although these knuckles are generally well fendered, the tug can be gainfully employed from the ships quarter, pulling into the tidal flow to ease the weight on the entrance knuckle.

The alternative position for a single tug could be to push on the vessels opposite quarter, into the direction of the tidal flow. This action would also ease the ships weight off the concrete knuckle. It is, however, pointed out that in the event of forward moorings parting, the tug would be at some risk in a position between the ship and the quayside in the pushing mode and may be exposed to an element of crushing action, if badly positioned.

Where two tugs are engaged, it would be normal practice for a conventionally sized vessel to secure one tug aft and one tug forward. The forward tug normally goes into the dock with the parent vessel. Once inside the dock entrance, the need and use of the aft tug is often dispensed with, having rounded the knuckle entrance.

> ***Comment***: Such manoeuvres are advised for the average size vessel, say 10,000 grt. Clearly super-tankers and the very large class of vessels would normally engage several tugs, minimum four, for most manoeuvring aspects.

Turning a vessel (with tug assistance)

Turning the vessel into a dock is achieved by first stemming the tide usually by turning short round or snubbing round on the ship's anchor, assuming that the vessel initially has the tidal/current astern.

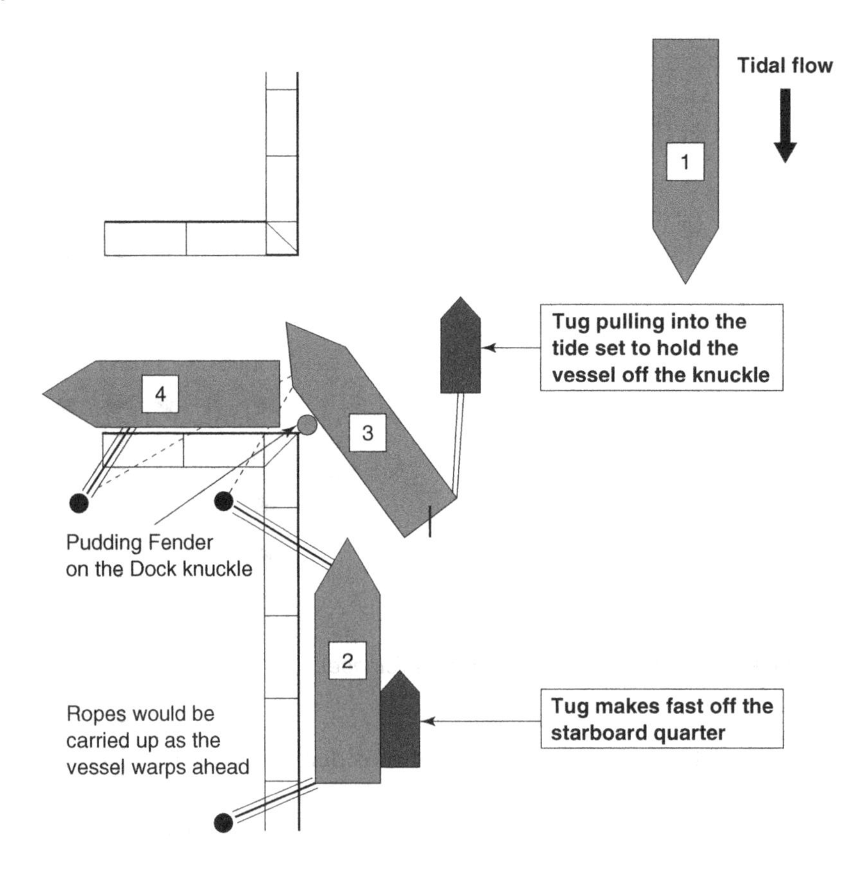

The position of the vessel '1' is seen as in the exposed water area with the tide astern.

Once the vessel has turned to stem the tide, position '2' is achieved by laying alongside the berth below the dock entrance. This position is held with mooring lines to prevent the vessel ranging on the quay. A breast line aft, with two head lines would be normal practice. The tug would normally make fast on the offshore quarter at this stage. Once the vessel starts to warp ahead the tug will pull from the offshore quarter against the tidal set, holding the vessel off the knuckle.

An alternative position for the tug to take up would be in a pushing mode off the inshore quarter, but this is not a position which is always favoured. In both positions the tug acts in opposition to the direction of the tidal stream.

Entering a dock (with alternative tug assistance)

Entering a dock with the aid of the assistance of a tug can be achieved by employing the tug either on the starboard quarter to pull against the tidal effect or to engage the tug pushing on the vessel's port quarter.

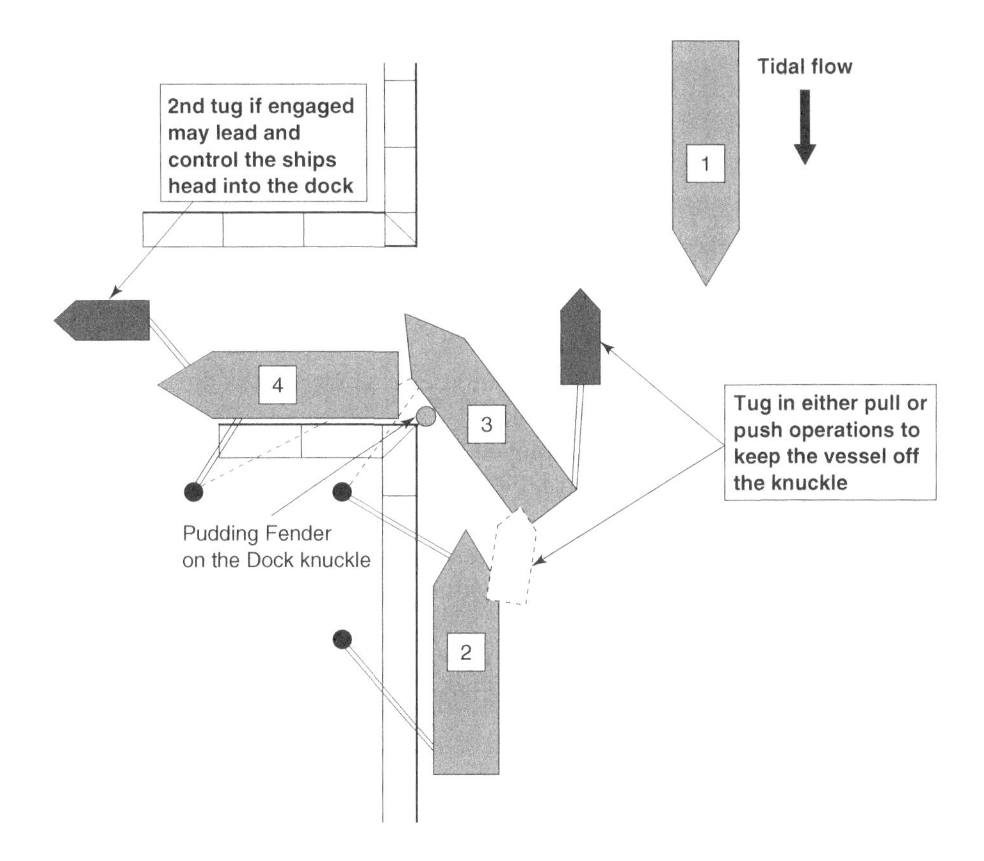

Tug use is favoured pulling on the starboard quarter position, rather than pushing on the port quarter. The reason for this is that the possibility of a head line parting or being otherwise lost at the bow position, could cause the mother vessel to fall back with the possibility of crushing the tug against the quayside wall.

Entering a 'dry dock' to take centre line blocks

The objective is to enter the dry dock from the tidal waterway and line the vessel up with the centre line blocks of the prepared dock.

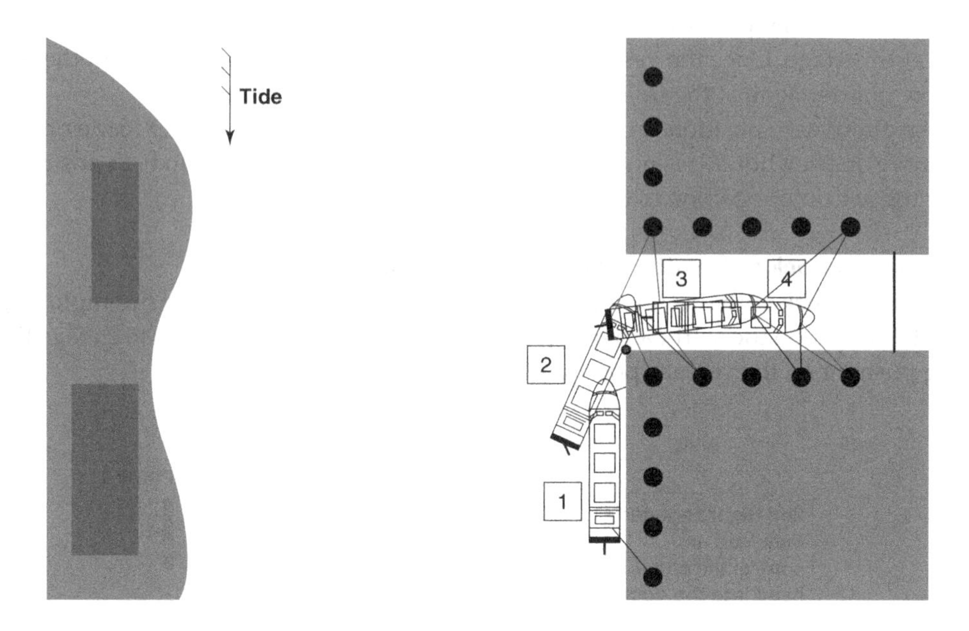

This manoeuvre is applicable to single ship docking operations, as with single ship graving docks or synchro-lift.

1. The vessel would stem the tidal flow and berth alongside the dock entrance in a similar manner as for entering a dock.
2. A pudding fender would be employed to protect the shell plate from landing heavily on the knuckle entrance of the dock, while moorings would be employed to warp the vessel into the dock entrance, by carry up methods.
3. In order to assist line up of the vessel, in a central dock position it would be considered essential to stretch moorings to either side of the docking area, in both for and aft positions.
4. Once the stern of the vessel has cleared the caisson (dock gate) astern and the gates are closed, alignment can take place with the advice from the dry dock personnel.

Aspects of dry docking

The vessel 'Carantec' lies in the floating dry dock in the Port of Barcelona, Spain. The empty graving dock lies to the right of the floating dock. The caisson is seen closed facing the harbour entrance.

The 'Carantec' seen from the seaward aspect, lying in the floating dry dock. The floating dock is semi-permanently moored in the wet basin, allowing access from the inner harbour waters.

Deck preparations – prior to berthing

It is general practice to call the ship's crew to their mooring stations in ample time to prepare the mooring deck for securing the vessel in a safe manner. The preparation tasks vary with the type of equipment aboard the vessel and the respective positions of the mooring deck(s). However, many activities are common to both the fore end and the after end of the vessel and will include the following:

1. All deck machinery by way of windlass, capstans, docking winches and powered roller leads should be tested and turned over and seen to be in working order.
2. All mooring ropes and wires should be flaked out clear of stowage reels and made ready for running ashore.
3. An adequate number of heaving lines of sufficient length should be available.
4. Sufficient fenders should be strategically secured overside to prevent bad landings likely to cause damage to the shell plate.
5. Anchors should be cleared away and left at a state of readiness for emergency use. This should be completed prior to the vessels approach to the channel or fairway, with anchors being left on the brake.
6. Communications between the bridge and mooring stations should be tested together with ship's external communication systems to tugs or harbour control stations.
7. Crew members should be retained to secure tug's lines if required.
8. Station Officers should inspect their respective mooring decks to ensure that a safe mooring operation can be carried out (any defects in equipment or short comings amongst crew members, should be reported to the ship's Master prior to closing the berth).
9. Specific equipment, like stoppers or messenger lines, should be readily available.
10. Relevant entries in the deck logbook should reflect the mooring deck preparations.

All participating crew members should be suitably dressed and protected to carry out their respective mooring duties.

Use of ship's lines

A passenger vessel is warped ahead from the forward mooring deck. An inshore and offshore mooring rope are engaged from mooring winch drums positioned either side, inboard of the split cable lifter, windlass arrangements. The mooring deck is liberally fitted out with bitts, roller fairleads, old men leads and panama pipe leads set into the bulwarks. The spare anchor is seen sited on the fore and aft line, between the two mooring winches.

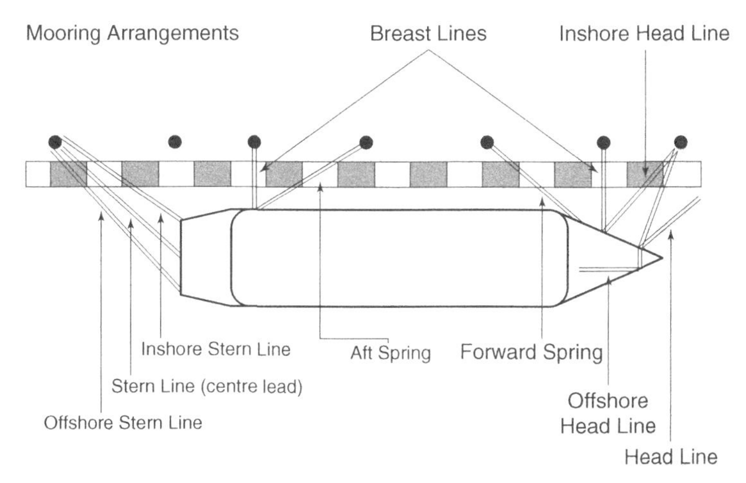

Mooring arrangements.

Passenger vessels 'Seawing' and 'Salamis Glory' secured by head lines and springs.

Mooring lines and deck equipment

The forward mooring deck

Conventional construction tends to favour open spaced mooring deck arrangements but this is no longer common practice across the varied range of specialized vessels. Many passenger vessels for instance which have high freeboards often have their mooring decks under cover, below a passenger foredeck. Similarly, some high speed ferries, with the catamaran or tri-maran hulls feature mooring arrangements undercover of upper decking. While some Ro–Ro, car carrier designs, because of the high central position of the stern vehicle ramp, have mooring decks established below decks and positioned to run moorings from the ship's quarters.

Conventional mooring operations

Conventional RoPax ferry operating with an exposed forward mooring deck, as seen from the Port Bridge wing. Bulwarks are fitted all round to the fore deck. A passenger promenade deck fitted with railings, is seen set below the Navigation Bridge Deck. The design does not include a bow visor arrangement.

Electric powered, split windlass/cable lifter, geared with a single shaft. The cable gypsy can be de-clutched to permit separate operations for the warping drum and for the mooring rope winch. The rope drum is purposely designed for the winch to have the brake applied, so relieving the need to transfer the mooring ropes onto 'bitts'.

Use of storm moorings

Where vessels lay overnight in order to maintain regular schedules, or when ships are alongside for long periods, additional storm moorings are often the order of the day. The concern is that high-sided ferries, vehicle carriers, bulk carriers when light, and large passenger ships, coupled with the danger of a high spring tide, will often present increased windage areas and become susceptible to being blown off the berth.

To counter this threat, several ports have established heavy duty combination moorings of nylon spring/coir rope, wire pennant construction, in addition to normal day-to-day mooring options, while the ship itself may employ additional security measures by deploying the insurance wire (if carried). Where insurance wires are used, these are usually turned up over at least two separate sets of bitts. The 58 mm wire is difficult to handle and turn up. Neither will it stretch when under tension, but it may just hold the vessel in position in the event of bad weather.

Where storm moorings are the order of the day, it should be remembered well before departure time that they are heavy to manhandle and as such, slow to land. Ample time should be allowed before sailing to clear these types of moorings prior to singling up on general mooring arrangements. Similarly, where specific storm moorings have not been deployed but additional moorings might have been ranged instead, Masters could expect to be engaged in more lengthy periods of departures and should allow greater time for singling up.

Bad weather is always problematical for high-sided vessels and their respective cargoes. The alternative to staying alongside during bad weather periods is to put to sea and take shelter in the 'lee' of a land mass and wait until the weather abates. The weather conditions would have to be extreme for the vessel to return to sea without completing cargo operations. However, a Master should choose the safest environment to place his vessel, and if the position alongside is compromised then choosing the open sea option rather than staying berthed and taking the force of, say, a Tropical Revolving Storm (TRS) may be preferable.

Whenever bad weather is experienced, regular monitoring of the weather forecast and plotting of the storm's position should be considered as an essential element of standing orders and good ship keeping. Passenger vessels clearly have sailing schedules to maintain but to this end they usually have adequate reserve power to engage when on route in the event of sailing times being delayed. The general public also accept that weather conditions may restrict operations and would not expect Masters or Companies to surrender safety, just to be expedient with its sailing schedule.

Mooring deck

The aft mooring deck of the vessel 'Andra'. Mooring ropes seen stretched from panama leads from each quarter and belayed to suitably sited sets of bitts. Double roller leads are also featured positioned on top of the bulwarks. A centre line capstan is actively employed instead of a docking/mooring winch. The motor for capstans is usually mounted below deck and the capstan is turned about a centre rotary axle.

Mooring equipment
Types of roller leads

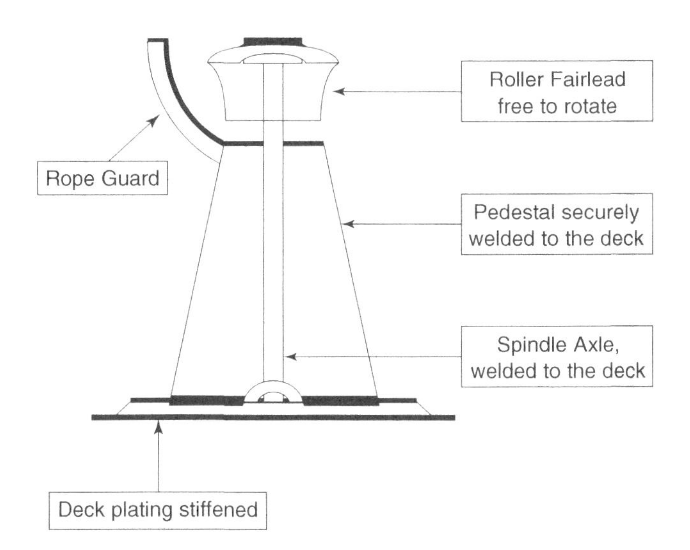

Mooring equipment: Types of roller leads. Pedestal roller fairlead welded to aft mooring deck. Colloquially referred to as an 'old man'.

Roller fairleads

Roller fairleads are generally popular in all sectors of commercial shipping. However, they can seize up unless regularly greased. Should a roller become 'frozen' it can usually be freed by a mooring rope generating a friction drive. Nevertheless, a good maintenance schedule should avoid this.

Roller fairleads come as single rollers, as an 'Old Man' on a free-standing pedestal, or as double or triple sets. The disadvantage with them is that moorings have been known to jump clear of the roller. Should such an incident occur as may happen with open triple sets constructed at the upper edge of bulwarks, the possibility of an accident becomes a reality.

NB. *This is the main reason why seamen tend to favour 'Closed Leads'.*

A double roller fairlead set into a side bulwark which by construction would prevent the mooring from accidentally jumping free. Roller leads avoid the sharp nip on the mooring and reduce the overall friction within the mooring. Such a reduction lends to easier handling when engaging moorings on warping drums.

International roller fairlead

The international roller fairlead is a popular multi-angled lead which allows moorings to be controlled at acute angles when the ship is rising or being lowered as when inside locks. The main disadvantage is that being rollers set about axles, they require regular greasing as part of the ship's planned maintenance. To this end, each roller is end fitted with a grease nipple.

This type of lead is generally found on the aft quarters at the stern mooring deck. They permit acceptable leads for mooring lines across wide transom sterns and

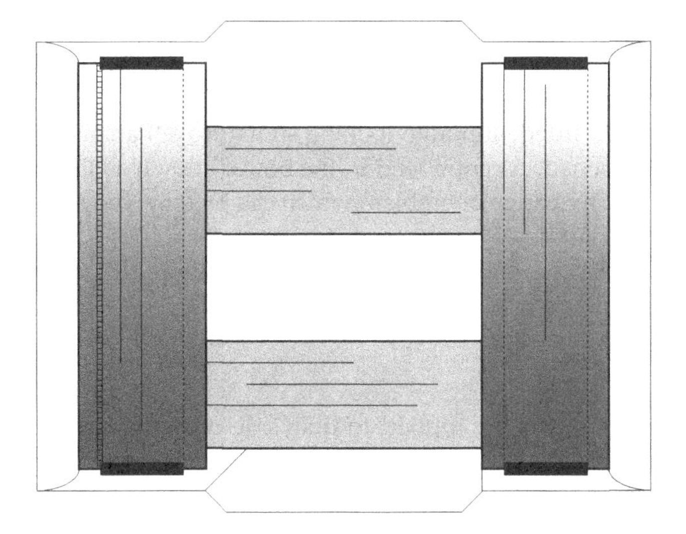

allow the rollers to accommodate wide angles to the quayside. They are not common to the bow region except possibly at the aft end of the forward mooring deck, in the region of the shoulder, where they can be used to run spring moorings.

Panama leads

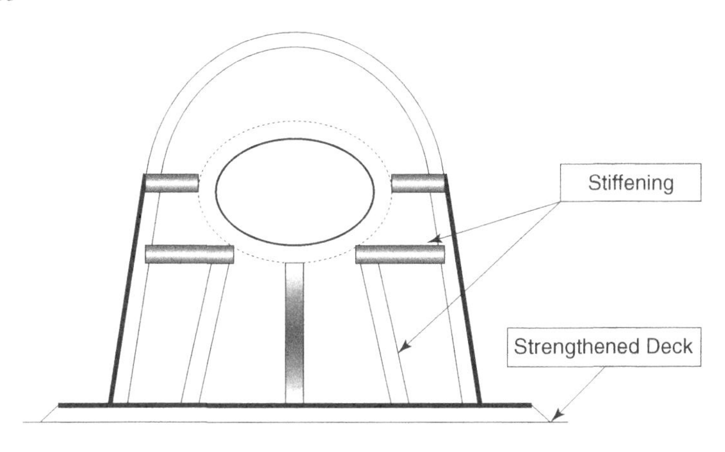

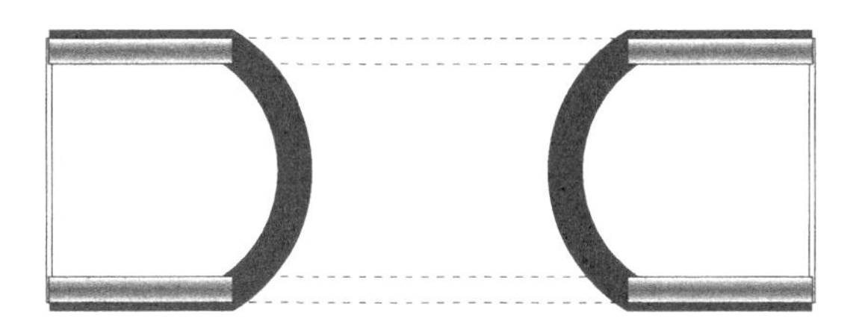

Panama leads, commonly referred to as 'pipe leads', are usually set into solid bulwarks with or without a doubling plate reinforcement. They are always well stiffened by angle bar supports.

The centre lead, if fitted, although looking similar to a 'Panama lead', is considerably more strengthened. A centre lead in the bows, known as the 'bullring' may be fitted with a company badge or emblem which can be removed for centre lead activity if and when required.

Also employed for the compulsory, emergency towing lead, which is a stipulated requirement for tankers.

Bollards (bitts)

The term 'bollard' is usually applied to quayside mooring posts. The term 'bitts' tends to refer to a double post bollard set, as mounted on the deck of virtually all ocean-going vessels. The securing of 'bitts' to the deck is more often achieved by a welded structure, although alternative methods have been employed in the past.

The posts themselves will be manufactured in cast steel or, more commonly, strengthened tubular steel. They are normally fitted with lugs and/or lips to the upper edge of the post to prevent the moorings jumping from the posts when under tension.

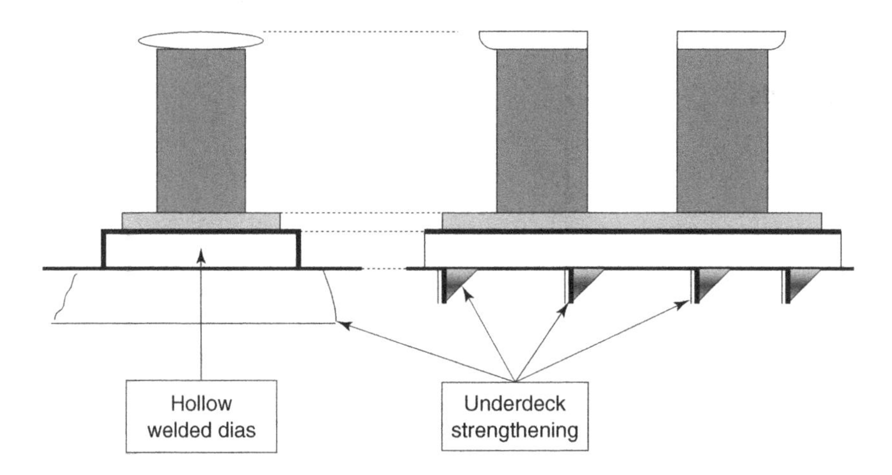

Double sets of bitts set on board the mooring deck of an offshore 'Anchor Handling' vessel. A stopper wire is seen being laid in 'Figure 8s' about the centre pair, the upper turns being lashed to prevent the wire springing off the 'Bitts' when under tension.

2 Manoeuvring characteristics and interaction

Introduction; Ship's performance factors; Pivot point action; Turning circles; Propeller action, transverse thrust; Angle of pitch, propeller slip; Twin screw vessels; Pod propulsion; High speed craft reactions; Wheel over points; Squat and blockage factor; Interaction scenarios.

Introduction

Each ship will have its own manoeuvring characteristics. The position of the pivot point will vary performance, while performance itself can be affected by numerous factors; not least, growth on the hull. The propellers, of such varied construction these days, can expect to generate increased thrust with reduced cavitation, while 'slip' and transverse thrust affects have as yet, not been eliminated from propeller activity.

Interaction inside the marine environment is noticeable in several forms, where a ship can experience a reaction from a land mass or another ship; typically, a parent vessel reacting with the smaller tug – the weaker element with the stronger. Interaction can be observed as squat, a bank cushion affect, or just an unexpected movement between two vessels in close proximity.

Whatever form interaction takes, it is generally seen as undesirable and unwanted. Mariners have become familiar with its effects over the years and the industry has gone some way to educate our seamen in anticipation of what to expect. Bearing this in mind, it would seem obvious to avoid the experience if possible, or if it is going to be encountered, then we should know how to counter its adverse effects.

Many factors are associated with interaction, not least speed of the vessel, depth of water, proximity of obstructions, the hull form and the manoeuvring aids operational to the vessel. Some influences can be avoided or even eliminated with awareness and training, while improved navigation practice must be expected to lessen the dangers and make ship handling safer to the individual and better for the environment.

The form of the land and the lack of underkeel clearance when vessels enter shallows will always be features worthy of special attention by the navigator and a ship's Master. Pilotage will always be directly affected by the shallow water effect, while overall performance must encompass the elements derived from propeller action and the combined effects of the environment on the hull.

See Appendix B for the Maritime Coastguard Agency's Marine Guidance Note on Dangers of Interaction – MGN 199.

A ship's performance factors

Ships are expected to meet their design speeds, and the propulsion units can only deliver the speed element provided all other related factors are also in place. Such elements as hull growth, corrosion or damage to a hull will clearly affect overall performance.

Corrosion

When a vessel is engaged in the regular employment of loading and unloading cargo, the ship's hull is continuously in and out of the water between the load and light ship draughts. Such movement makes the hull form, susceptible to corrosive effects and allows rust to develop. As corrosive layers come away from the paint film, the hull is left indented and some parts are left proud. Such unevenness can generate hull resistance until overside maintenance can be applied – usually when the ship is in dry dock.

Hull growth

Where a vessel is frequently in port or operating in a river, it is highly probable that the hull will attract weed and similar organic growth. This provides additional resistance to the hull's movement through the water and, unless regularly cleaned off in dry dock, could eventually cause a reduction in speed performance. This growth may also include barnacles attaching to the hull and to the propeller(s) – causing additional resistance which would affect propeller rotation speed and subsequent fuel burn.

Indentation hull damage

During a ship's life, landing and berthing alongside docks and quaysides tend to take a toll on the smoothness of the hull's lines. Such damage will affect the water flow around the hull, and further resistance to passage through the water will be encountered. Although seemingly minor at the time, this type of hull damage can and does accumulate over a period of time, which can again directly affect the ship's performance.

Engine maintenance

Clearly it is well recognized that you cannot get out what you do not put in. An engine will only deliver peak performance from continuous high standards of maintenance. Fuel quality is also critical to the machinery output and subsequent propeller performance. Therefore, plant needs to be operated under a planned maintenance schedule where all elements are monitored on a regular basis; effective performance of machinery being linked directly to effective maintenance.

Manoeuvring information

It is now recommended that manoeuvring information in the form of a 'Pilot Card', 'Bridge Poster' and 'manoeuvring booklet' should be retained on board ships. Such information should include comprehensive details on the following factors affecting the details of the ship's manoeuvrability, as obtained from construction plans, trials and calculated estimates.

Ships general particulars – Inclusive of name, year of build and distinctive identification numbers; gross tonnage, deadweight, and displacement at summer draught; the principle dimensions, length overall, moulded breadth and depth, summer draught and ballast draught and the extreme height of the ship's structure above the keel.

Listed main manoeuvring features – Main engine, type and number of units, together with power output; the number and type of propellers, their diameter, pitch and direction of rotation; the type and number of rudders with their respective areas; bow and stern thruster units (if fitted), type and capacity.

Hull particulars – Profiles of the bow and stern sections of the vessel and the length of the parallel of the middle body (respective to berthing alongside).

Manoeuvring characteristics in deep and shallow waters – Curves should be constructed for shallow and restricted waters to show the maximum squat values at different speeds and blockage factors, with the ship at variable draughts.

Main engine – Manoeuvring speed tables established for loaded and ballast conditions from trials or estimated; stated critical revolutions and maximum/minimum revolutions; time periods to effect engine telegraph changes for emergency and routine operational needs.

Wind forces and drift effects – The ability of the ship to maintain course headings under relative wind speeds, should also be noted; together with the drifting effects on the vessel under the influence of wind, when the vessel is without engine power.

Manoeuvring characteristics in deep water

Course change performance – Turning circle information from trials or estimates for various loaded/ballast conditions; Test condition results reflecting 'advance' and 'transfer' and the stated maximum rudder angle employed in the test, together with times and speeds at 90°, 180°, 270° and 360°; details should be in diagrammatic format with ship's outline.

Acceleration and speed characteristics – Presentation of speed performance when the ship accelerates from a stopped position and deceleration from full sea speed to a position of rest, reflecting maximum rudder angles, for loaded and ballast conditions.

Stopping capabilities – Should include respective track stopping distances from:

Full astern from a position of full ahead sea speed

Full astern from a position of full ahead manoeuvring speed

Full astern from half ahead

Full astern from slow ahead

Stopping the engine from a position of full sea speed ahead

Stopping from a position of full manoeuvring speed ahead

Stopping engine from half ahead

Stopping engine from slow ahead.

Relevant time intervals should also be recorded, reflecting the time to reach full ahead and positions of zero speeds, compatible with the above operations.

Information on the minimum speed (rpm) that the ship can retain steerage capability.

Any other relevant information considered useful to the manoeuvring and handling capabilities of the vessel should be included in this 'Manoeuvring Booklet'.

The ship's pivot point

The turning effect of a vessel will take effect about the ship's 'pivot point' and this position, with the average design vessel, lies at about the ship's Centre of Gravity, which is generally nearly amidships (assuming the vessel is on even keel in calm water conditions).

As the ship moves forward under engine power, the pivot point will be caused to move forward with the momentum on the vessel. If the water does not exert resistance on the hull the pivot point would assume a position in the bow region. However, practically the pivot point moves to a position approximately 0.25 of the ships length (L) from the forward position.

Similarly, if the vessel is moved astern, the stern motion would cause the Pivot Point to move aft and adopt a new position approximately 0.25 of the ship's length from the right aft position.

If the turning motion of the vessel is considered, with use of the rudder, while the vessel is moved ahead by engines, it can be seen that the pivot point will follow the arc of the turn.

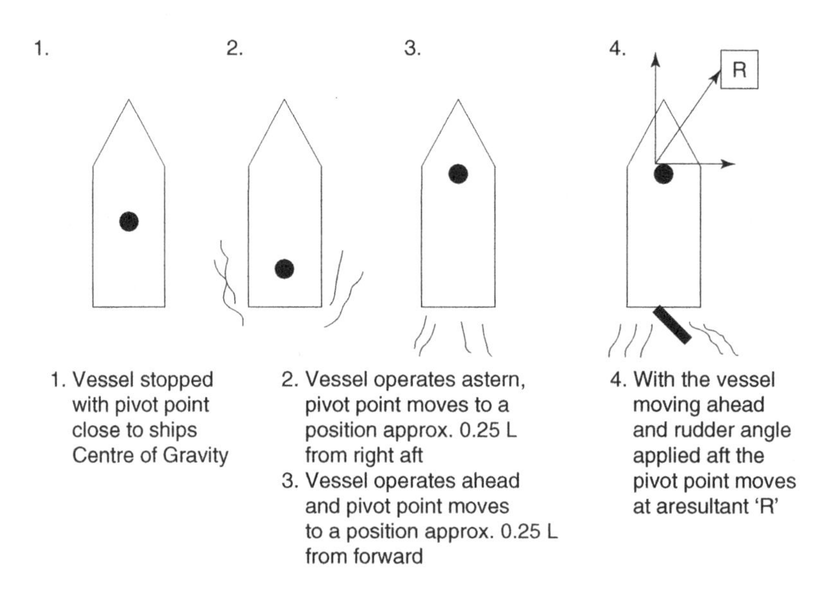

1. Vessel stopped with pivot point close to ships Centre of Gravity
2. Vessel operates astern, pivot point moves to a position approx. 0.25 L from right aft
3. Vessel operates ahead and pivot point moves to a position approx. 0.25 L from forward
4. With the vessel moving ahead and rudder angle applied aft the pivot point moves at aresultant 'R'

When the vessel is moving ahead and turning at the same time, the forces on the ship take affect either side of the pivot point, as shown below:

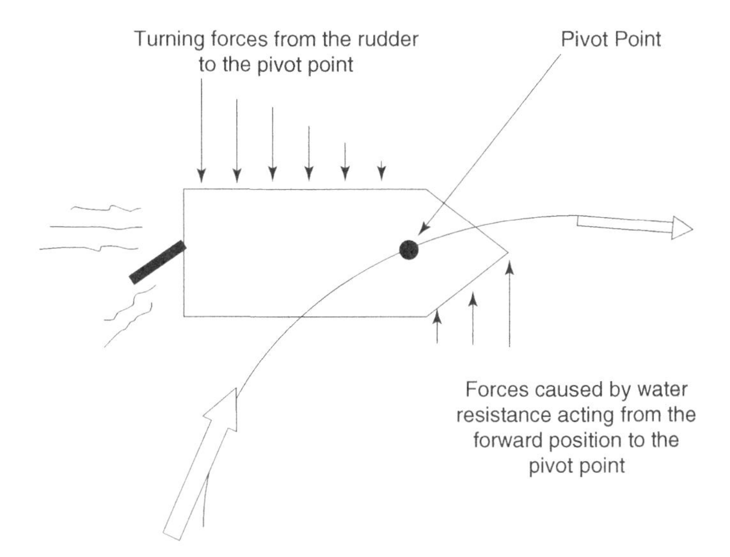

The combined forces of water resistance, forward of the pivot point and the opposing turning forces from the rudder, aft of the pivot point, cause a 'couple effect' to take place. The resultant turning motion on the vessel sees the pivot point following the arc of the turn.

The pivot point at anchor

It should be noted that when the vessel goes to anchor the pivot point moves right forward and effectively holds the bow in one position. Any forces acting on the hull, such as from wind or currents, would cause the vessel to move about the hawse pipe position.

Use of the rudder can, however, be employed when at anchor, to provide a 'sheer' to the vessel, which could be a useful action to angle the length of the vessel away from localized dangers.

Turning circles and advice on turning

Turning circles are normally carried out during the sea trials of the vessel prior to handover from builders to owners. The fact that the manoeuvre may have to be carried out at sea, for collision avoidance purposes, makes this an item of 'need to know' for the ship's Master and Watch Officers.

The ship's trial papers and performance criteria will be placed on board the vessel prior to handover. Statements as to the 'advance' of the vessel and its 'transfer' will be stated, together with the 'Tactical diameter' and 'Final diameter' that the vessel scribes on trials. It should be realized that trials are generally conducted in relatively calm weather conditions with little wind. In reality, should it become necessary to execute a 'round turn', conditions are unlikely to be the same and, therefore, the criteria provided will not necessarily be the same as that provided in trial documents.

In the turning circle example shown on page 39 of a Cargo/Passenger Ferry vessel, the given helm was 35° hard over to each side for the respective turns to port and to starboard. The measurements for the Tactical and final diameters are indicated, as is the transfer on each of the respective turns. The ship was fitted with triple Controllable Pitch Propellers and conducted the turns at 20.3 knots (starboard) and 20.2 knots (port) from diesel (Sultzer) engines delivering 8000 h.p.

Turning circle – definitions and features

Once trials of a new ship are complete, operators will need to know how the vessel can expect to perform in a variety of sea conditions. The ship handler, for instance, should be aware of how long it will take for a vessel to become stopped in the water from a full ahead position or how far the vessel will advance in a turn. Turning circles and stopping distance (speed trials) provides such essential information to those that control today's ships.

Advance – Defined by the forward motion of the ship, from the moment that the vessel commences the turn. It is the distance travelled by the vessel in the direction of the original course from commencing the turn to completing the turn. It is calibrated between the course heading when commencing the turn, to when the vessels head has passed through 90°.

Transfer – Defined by that distance which the vessel will move perpendicular to the fore and aft line from the commencement of the turn. The total transfer experienced during a turn will be reflected when the ship's head has moved through a course heading of 180°. The amount of transfer can be calibrated against the ship's change of heading and is usually noted at 90° and 180°.

Tactical diameter – Is defined by the greatest diameter scribed by the vessel from commencing the turn to completing the turn.

Final diameter – Is defined as the internal diameter of the turning circle where no allowance has been made for the decreasing curvature as experienced with the tactical diameter.

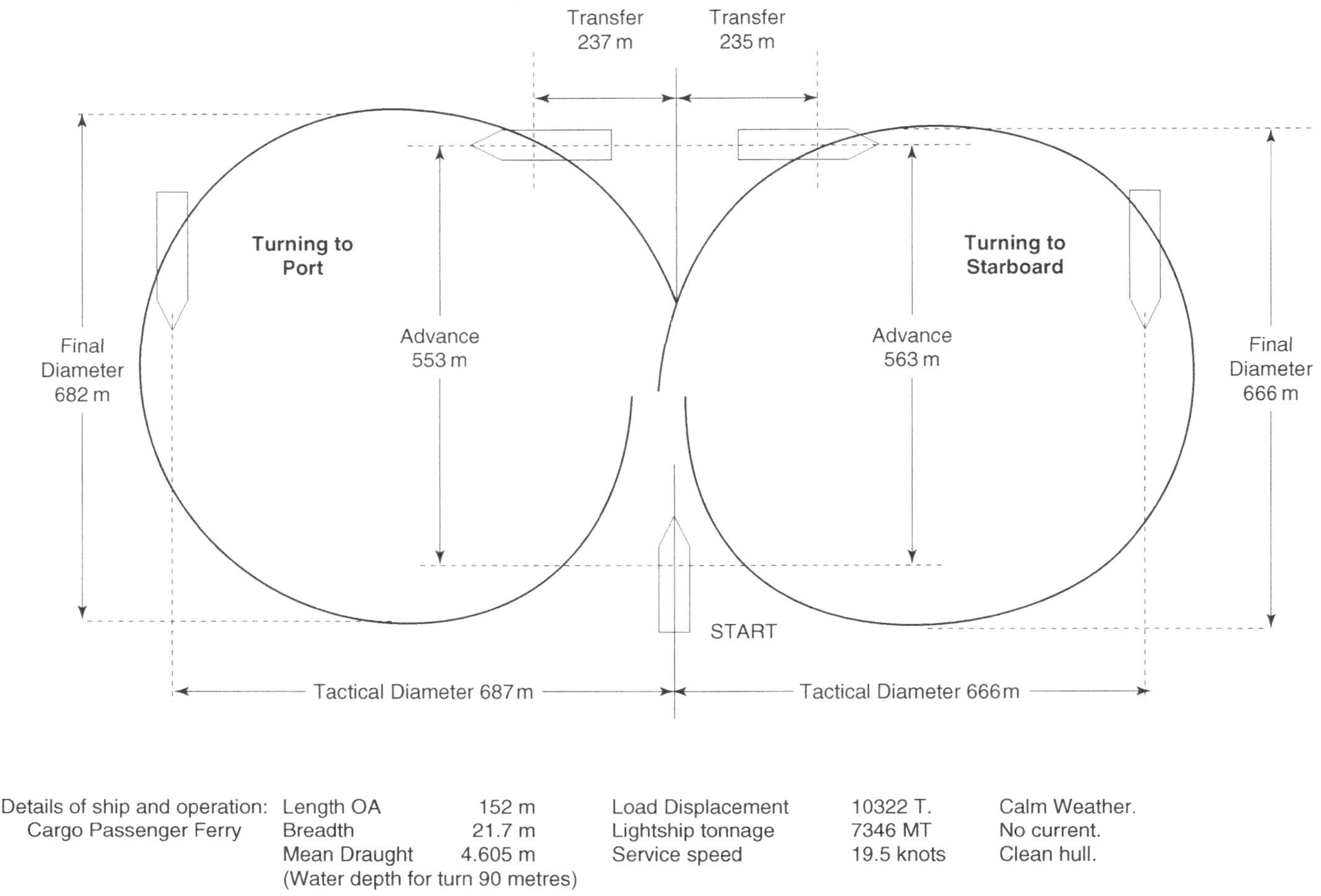
Transfer
237 m
Transfer
235 m
Turning to
Port
Turning to
Starboard
Final
Diameter
682 m
Advance
553 m
Advance
563 m
Final
Diameter
666 m
START
Tactical Diameter 687 m
Tactical Diameter 666 m
Details of ship and operation:
Cargo Passenger Ferry
Length OA 152 m
Breadth 21.7 m
Mean Draught 4.605 m
(Water depth for turn 90 metres)
Load Displacement 10322 T.
Lightship tonnage 7346 MT
Service speed 19.5 knots
Calm Weather.
No current.
Clean hull.

General information on turning circles

The conditions prevailing during the turning of a vessel will greatly affect the determined results. A major example of this would be experienced where a circle is considerably increased in size when conducted in very shallow waters, especially when compared with a turn conducted in deep waters. It would, therefore, be fair to assume that the turning rate of the quickest turn might not generate the tightest of turns.

It is also noted that the action of turning the vessel with hard over helm on, would cause the ship's speed to decrease by a considerable amount. A drop of 30 to 40 per cent from full speed would not be seen as unexpected, assuming no direct reduction to the propulsion unit is applied. The rudder angle imposed, generating considerable drag effect during the turn, accounts for some loss of speed while the fore and aft component of hydrodynamic forces also cause a speed reducing affect, slowing the vessel down during the turn.

When conducting turns at high speed the only thing that is saved, is time, while the 'rate of turn' varies considerably. Such a factor may be critical in certain cases, especially where time is the important factor, as in the case of the man overboard situation.

Turning features – operational vessels

Once operational and a vessel has reason to perform a tight turn, e.g. Man Overboard, it should be realized that a deep laden vessel will experience little effect from wind or sea conditions, while a vessel in a light ballast condition, may experience considerable leeway with strong winds prevailing.

Another feature exists with a vessel that is trimmed by the stern. She will generally steer more easily, but the tactical diameter of a turn could be expected to decrease; while a vessel trimmed by the head will still decrease the size of the circle, but will be more difficult to steer.

Should the vessel be carrying a list at the time of conducting the circle, the completion time could expect to be delayed. Also, turning towards the list would expect to generate a larger turning circle than turning away from the list side, bearing in mind that a vessel tends to heel in towards the direction of the turn, once helm is applied.

It should also be realized that a ship turning with an existing list and not in an upright condition, especially in a shallow depth, could experience an increase in draught. Such a situation could also result in reduced buoyancy under the low side causing a degree of sinkage to take place. This increase in draught would not be enhanced if the turning action was also being conducted at high speed.

Additional considerations

The features associated with turning a vessel will be influenced by the type of rudder employed with the ship. This could be readily accepted if a conventional semi-balanced bolt axle rudder is considered against, say, a flap design rudder which would generate a substantially greater turning lever, producing a greatly reduced turning circle.

A narrow beam vessel like a warship, would also tend to make a tighter turning circle than a wide beam container vessel. So, respectively, the construction of the hull, the manoeuvring equipment together with speed of turn, draught, geographic water conditions, state of equilibrium are all relevant and must all be seen as influential factors relating to the effective turning of the vessel.

Influences on the turning circle

Modern day ships are built with a variety of manoeuvring aids. The previous example is unusual in that it had triple controllable pitch propellers. However, many ships are still being constructed with a righthand fixed propeller. Generally speaking, such vessels would turn tighter to port than to starboard, although weather conditions on the day of trials could influence this. Other factors will affect the rate of turn and size of the actual circle, namely:

a) Structural design and length of the vessel
b) Draught and trim of the vessel at the time of trials
c) The size and motive power of machinery employed
d) Distribution and stowage of any cargo
e) Whether the ship is on even keel or carrying a list
f) The geographic position of the turn and the available depth of water
g) The amount of rudder angle applied to complete the turn
h) External forces effecting the drift angle.

Structural design and ship's length

Generally speaking, the longer the ship, the greater the turning circle. The type and surface area of the rudder will also have a major influence in defining the final diameter of the circle, especially the clearance between the rudder and the hull. The smaller the clearance between the rudder and the hull, the more effective will be the turning action.

Draught and trim

The deeper a vessel lies in the water the more sluggish will be her response to the helm. However, where a vessel is in a light condition and at a shallow draught then the superstructure is more exposed and would be more influenced by the wind. The trim of the vessel will influence the size of the circle considerably. Ships usually trim by the stern for ease of handling purpose but it should be noted that if the vessel was trimmed by the head during the turn the circle would be distinctly reduced.

Motive power

The relationship between power and the ship's displacement will affect the turning circle and can be compared with a light-speed boat against a heavy ore carrier; the acceleration of the light-speed boat achieving greater manoeuvrability. Also, for the rudder to be effective, it must have a flow of water passing it. Therefore, the turning circle will not be increased by a great margin with an increase in speed because the steering effect is increased over the same common period.

Distribution and stowage of cargo

Ships' trials are generally conducted on new ships and cargo stowage on board is rarely a factor to consider. However, if cargo is on board the vessel would respond more favourably if the loads could be stowed in an amidships position as opposed to in the extremities of the vessel. Where loads are at the ends of the vessel, any manoeuvre would be sluggish and slow in response to helm action.

Even keel or listed over

It would normally be expected that a new vessel completing sea trials would be on an even keel throughout, but such a condition cannot be guaranteed once the ship is in active service. In the event that the vessel is carrying a list when involved in a turn she can be expected to make a larger turn when turning towards the side carrying the list. The opposite holds good when turning away from the listed over side and tends to make a reduced turning circle.

Available depth of water

Turning circles for trials should always be carried out in deep water. Shallow water would be expected to cause a form of interaction between the hull and the sea bed causing the vessel's head to yaw and it becomes more difficult to steer. Shallows could affect response time and so cause an increase in the advance and the transfer of the circle.

Rudder angle

A prominent feature of any turning operation and one where the optimum rudder angle is that which will cause a maximum turning affect with the reduced amount of drag. Where a large rudder angle is employed the turning circle would be tighter but it would be accompanied by a considerable loss of speed.

Drift angle and influence forces

When helm is applied and the bow responds, the stern of the vessel will traverse in an opposing direction. The resulting motion is one of a sideways movement at an angle of drift. When completing the turning circle, the stern of the vessel is outside the turning circle, while the bow area is inside the circle. In the majority of cases, it is the pivot point of the vessel which describes the perimeter of the turning circle.

Propeller action

Propellers are designed to produce maximum efficiency from the engine at the most economical fuel burn. However, the propeller itself gives rise to some drag effect and will have transverse thrust as a side effect. A degree of cavitation on the forward side of the blades can also be expected. Such effects continually reduce the propeller's effectiveness and have associated side effects like generating excessive vibration and noise.

The rotation of the propeller and the generation of cavitation leads to a vortex being created in the region of the blade tips. This influences the slip value and hence the speed of the vessel. This action could cause damage through 'pitting' which could also affect propeller performance.

There are now many different types of propeller systems in operation. The right hand fixed blade propeller is still common but developments in controllable pitch propellers, contra-rotating propellers, multi-blade propeller systems, twin, triple and quadruple propeller sets, pod propulsion units, Kort nozzle systems and Azipod systems, have all taken market share in both commercial and warship construction. Active rudders with propellers attached are also an added feature, while Voith Schneider Propellers (VSP) have made advances with the Voith Cycloidal Rudder working in conjunction with cycloidal propulsion.

Distinct advantages with each system are advocated by the various manufacturers, but generally the performance with respect to the vessel design and the designate vessel function, lend to a specific choice of system, e.g. Azipod systems for Dynamic Positioned vessels.

Transverse thrust

Transverse thrust effects are a cause of the single propeller action where water is displaced to one side or another, causing a movement of the hull from the deflection of the water flow. The effects of transverse thrust when going ahead are so minimal they can generally be ignored but when operating astern propulsion, the water flow expels water in the forward direction. This in turn is deflected by the hull form causing a sideways push on the hull.

The ship handler should be aware of his or her own vessel's performance when going astern and the diagram below goes some way to explaining the movement of the vessel with alternative propeller systems.

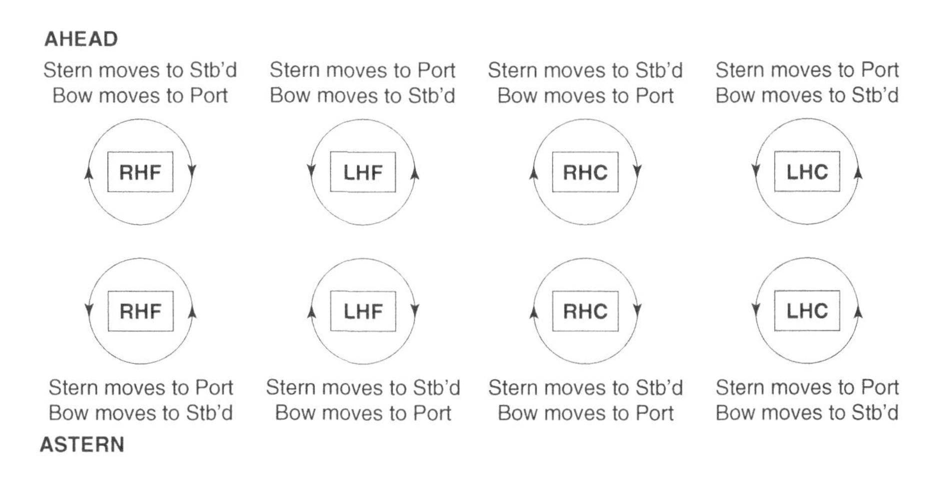

RHF, Right Hand Fixed Propeller; LHF, Left Hand Fixed Propeller; RHC, Right Hand Controllable; LHC, Left Hand Controllable.

Factors of propellers

Single fixed pitch propeller

Fixed pitch propeller(s) are subject to drag effects and slip, when the vessel is moving through the water. Being usually constructed in a dissimilar metal to the steelwork of the hull, they are subject to pitting and corrosion effects necessitating, in most cases, the use of sacrificial anodes about the rudder propeller area. These anodes can themselves generate some frictional resistance.

In the event of damage to one of the blades of the propeller, it would become necessary to replace the whole propeller. Changing a propeller is expensive and will usually require the vessel to enter dry dock.

NB. Historically smaller vessels could, and have been known to, change a propeller while alongside with the assistance of shoreside cranes and excessive forward trim.

Controllable pitch propellers (CPPs)

These are more expensive to fit than fixed pitch propellers, especially if they are to be fitted retrospective to the building stage. They are subject to more maintenance but have distinct advantages over and above fixed pitch blades. If a blade is damaged it can be removed comparably quickly and replaced by a spare blade (usually carried by the ship itself). The whole propeller does not need to be replaced.

The CPP is also cost-effective in that, with a constant rotating shaft, shaft alternators can be used for electrical power generation without having to resort to the use of additional generators. Additional generator units require expensive auxiliary fuel, a necessity with fixed pitch propellers.

The benefits to the ship handlers are immediate bridge response to ship control, without having to go through engineers to obtain manoeuvring controls. However, the controllable propellers still generate an element of drag effect, especially at zero pitch and they are also subject to similar corrosion as the fixed pitch propellers, for the same reasons. They are generally subject to reduced slip values.

Nozzle propellers

When operating at high speed, these propellers experience a reduced value of slip but when at slow speed, under heavy load, may experience increased values of slip. They also experience erosion on the inside edge of the nozzle and at the blade tips, usually due to cavitation and vortex effects. The nozzle itself tends to be protective and tends to prevent debris hitting the actual propeller blades.

CPP construction. The circular base of CPP blades bolted onto the rotational base, set into the propeller shaft of a Controllable Pitch Propeller arrangement. Once the bolts are secured, they are strap welded together to prevent loosening through vibration.

Steerage problems may also be experienced, especially where the nozzle is short in length compared with diameter. Some more recent nozzles are fitted with a steerage vane and alternative nozzle lengths may have been fitted as alternative options.

Pitch angle of propellers

Pitch – Is defined as the axial distance moved by the propeller in one revolution, through a solid medium.

Measurement of pitch

Most modern shipyards would establish the pitch of a propeller by the use of an instrument known as a 'Pitchometer'. However, if this was not available the pitch can be ascertained in dry dock from the exposed propeller.

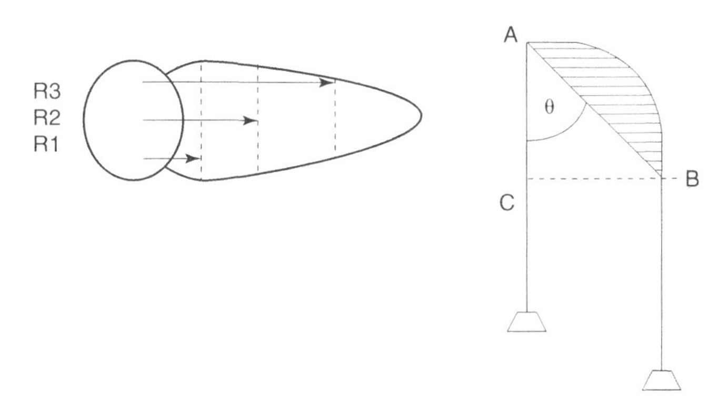

Position the propeller blade in the horizontal position and place a weighted cord over the blades.

At different Radi R1, R2, and R3, measure the distances AC and BC as well as R1, R2, and R3.

The tangent of the pitch angle is then $\frac{BC}{AC}$

$$\text{Therefore Pitch} = 2\pi \text{ R Tan } \theta$$

$$= \frac{2\pi \times R \times BC}{AC}$$

Example to calculate pitch angle of a propeller

In a four bladed propeller of constant varying pitch the following readings are obtained.

Pitch Angle	*Radi*
40°	0.5 m
25°	1.0 m
20°	1.5 m

Calculate the mean pitch of the blades.

$$\text{Pitch of propeller} = 2\pi \text{ R Tan } \theta$$
$$\text{At radi } 0.5 = 2\pi \times 0.5 \times \text{Tan } 40^\circ = 2.636$$
$$\text{At radi } 1.0 = 2\pi \times 1.0 \times \text{Tan } 25^\circ = 2.929$$
$$\text{At radi } 1.5 = 2\pi \times 1.5 \times \text{Tan } 20^\circ = 3.430$$
$$3\,\overline{)8.995}$$

Mean Pitch (Average) = 2.998 metres The theoretical distance the propellor will advance in one revolution.

Examples of propeller slip

Real slip – occurs as a result of physical conditions existing between the propeller and the water in which it is immersed. It should only be positive.

Apparent slip – is concerned with the same factors but in addition the effects of current and/or wind are taken into account. This may be positive or negative.

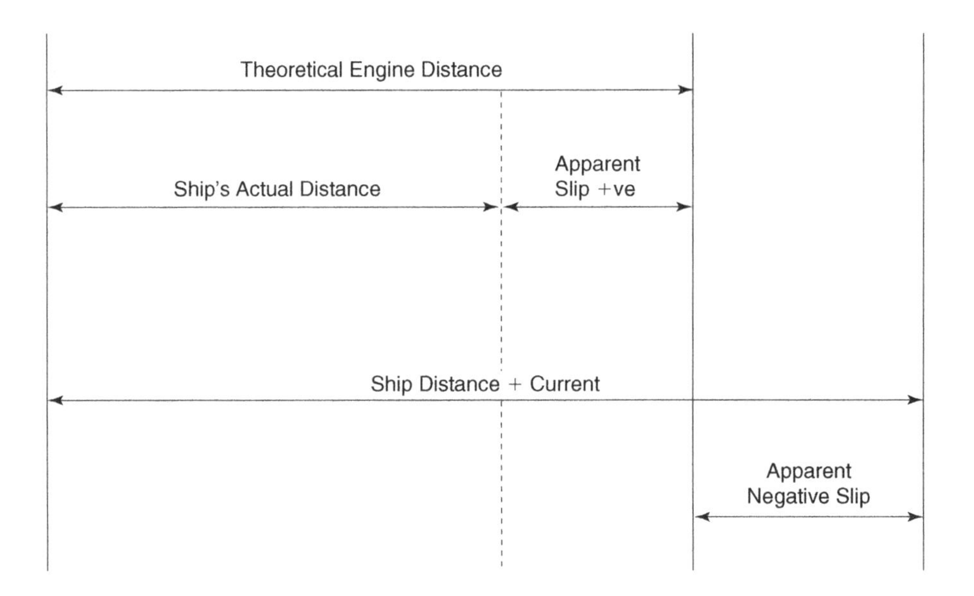

$$\% \text{ Slip} = \frac{\text{Engine Distance} - \text{Ship's Distance}}{\text{Engine Distance}}$$

Further example of propeller slip

Calculate the slip incurred by a vessel when given the R.P.M. = 125

And the pitch of the propeller is 6.0 m

The ship's run from Noon to Noon ship's mean time covers 540 nautical miles

Clocks are advanced 30 minutes during the day's run

$$\% \text{ Slip} = \frac{\text{Engine Speed} - \text{Ship's Speed}}{\text{Engine Speed}}$$

$$\text{Ship's speed} = \frac{\text{Distance}}{\text{Time}} = \frac{540}{23.5} = \underline{22.96 \text{ kts}}$$

$$\text{Propeller (Engine) Speed} = \frac{\text{Pitch} \times \text{RPM} \times 60}{1852}$$

$$= \frac{6 \times 125 \times 60}{1852}$$

$$= \underline{24.3 \text{ kts}}$$

$$\text{Slip} = \frac{24.3 - 22.96}{24.3} \times 100$$

$$= \underline{+5.5\%}$$

Twin screw vessels

It should be realized from the onset, that when dealing with twin screw vessels, some basic information is directly linked to the behaviour when manoeuvring the vessel:

a) Fixed propellers are usually both outward turning (some tonnage still has inward turning, fixed pitch propellers).
b) Controllable pitch propellers are usually inward turning.
c) Configurations can be twin rudders or a single rudder.
d) Twin rudder configurations are generally accepted as being more responsive than a single rudder with twin propellers.
e) Twin propellers with a single rudder configuration tend to have a poor water flow pattern over the rudder area, making the rudder less effective.
f) Where the propellers and twin rudders are inset close to the fore and aft line, the turning ability of the vessel is often reduced; the turning effect being insignificant in some narrow beam vessels, such as some classes of warships.
g) The effects of transverse thrust are still present and should be related as the same to a single fixed pitch propeller. Though the effects are considered a poor turning element if the vessel is having to manoeuvre in confined waters.
h) The rudder(s) turning force when the vessel is operating ahead propulsion is usually considered as very good.
i) The wash from propellers when the vessel is operating astern propulsion will tend not to extend up the hull length, if the vessel is at high speed.
j) Some twin screw ships will respond well when only one engine is operational with the twin rudders working in tandem. While other vessels could find that the non-operational engine shields its respective rudder making it less effective. Alternatively, the constructional hull lines of the vessel could well influence the flow to the rudder surface and directly effect the turning ability to a lesser or greater degree.

k) Shaft alignment of twin propellers has a direct influence on turning ability of the vessel. Parallel shafts (to the fore and aft line provide greater leverage about the pivot point. Angled shafts (slightly outboard), provide reduced leverage and subsequently reduced turning ability.
l) Twin screw vessels can be steered by engines with fine adjustments to the revolutions on respective shafts. Steerage would not be as accurate, or as steady as with rudder use, but would be manageable in an emergency.

Twin screw arrangements

Twin screw arrangements. Twin controllable pitch propellers (CPPs) each fitted with ducting and flap rudders, designed either side of a single skeg stern structure. The vessel is also fitted with twin stern thrusters set forward of the propellers, on the centre line above the keel. Extensive use of sacrificial anodes have been used to reduce the corrosive effects in the stern area due to the construction in dissimilar metals, namely with tail end shafts, bronze propellers and the steelwork of the rudders and hull.

Twin, four-bladed, fixed pitch propellers, positioned either side of a single balanced rudder. The arrangement seen fitted to the cable ship 'Nexus' exposed in dry dock.

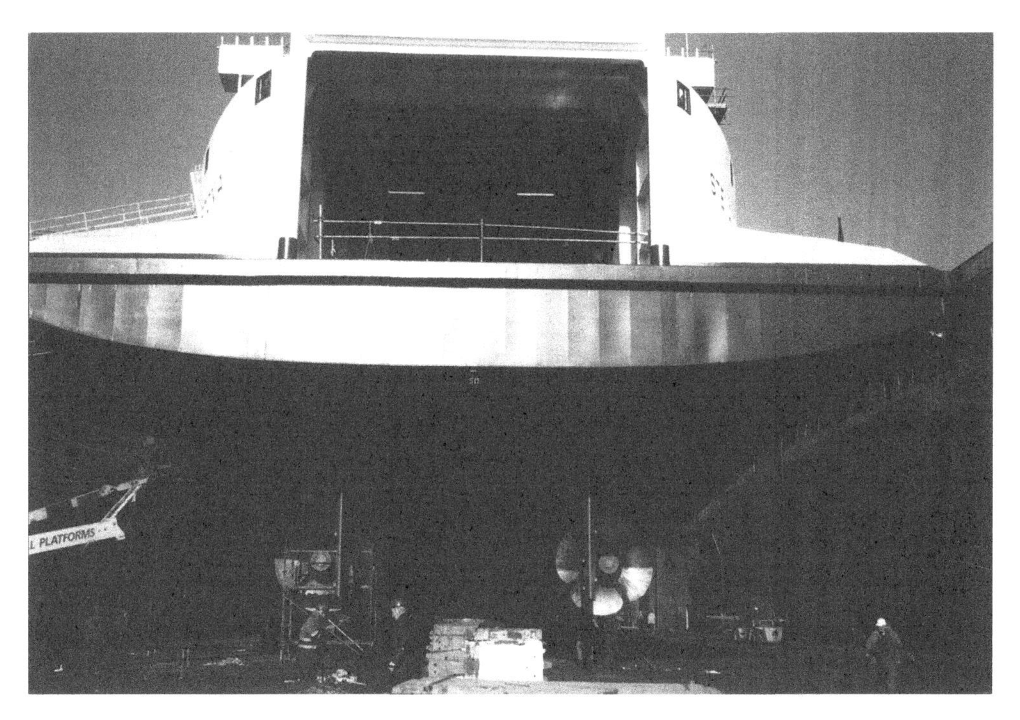

Twin multi-blade propellers fitted in association with twin rudders aboard a vehicle ferry, seen exposed in dry dock.

Machinery 'Pod' propulsion (Pod propulsion units)

Several cruise ships have recently moved towards 'Pod propulsion units' as a means of main power and many buildings in the ferry sector now reflect the potential use of 'Pod Technology' for vessels of the future. The compactness of the 'pod' and the associated benefits to passenger/ferry operations would seem to offer distinct advantages to ship handlers, operators and passengers alike.

Some of the possible advantages from this system would be in the form of:

1. Low noise levels and low vibration within the vessel.
2. Fuel efficiency with reduced emissions.
3. Good manoeuvring characteristics and tighter turning circle as when compared with a similar ship operating with standard shaft lines and rudders.
4. Reduced space occupied by bulky machinery making increased availability for additional freight or passenger accommodation.
5. Simpler maintenance operations for service or malfunction (pods are easy to remove/and replace).

Machinery pods are usually fitted to the hull form via an installation block, each vessel having customized units to satisfy the hydrodynamics and the propulsion parameters. Propeller size and the rpm would also need to reflect the propulsion requirements to the generator size with electric 'Azipod Units'.

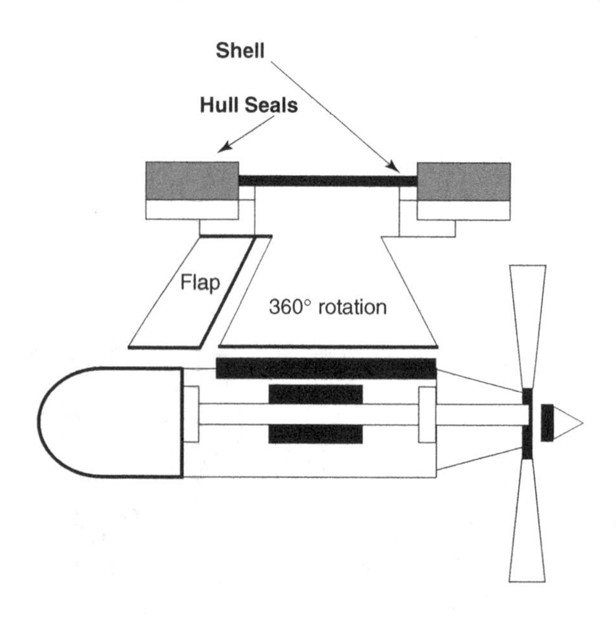

The direction of the shaft line is acquired from a hydraulic steering unit giving the versatility of directional thrust to port and starboard as well as ahead or astern.

For extremely high speed steering a 360° rotation pulling pod with a rudder flap has been designed.

Control means is provided by flap movement with the complete 'Pod' turning.

Azipod propulsion systems provide the action of pulling, rather than pushing the vessel through the water. A typical twin propeller azipod configuration would consist of three main diesel generators driving an electric motor to each propeller, with full bridge control transmission. Power ranges start from about 5 MW up to 38 MW dependent upon selected rpm (adequate built-in redundancy is accounted for by providing three generators for only two propellers).

The Azipod propulsion system makes ship handling easier and turning circles are comparatively tighter than where vessels are fitted with conventional rudders – speeds of 25 knots ahead, 17 knots astern and 5 knots sideways provides excellent harbour manoeuvring. Varieties of pod designs are rapidly entering the commercial market supported by associated new ideas to improve fuel efficiency and provide better performance.

Many are water cooled, eliminating the need for complex air cooled systems, while the Siemans-Schottel Propulsion (SSP) system has propellers at each end of the pod rotating in the same direction.

The Passenger vessel 'Amsterdam' fitted with twin azipod propeller units either side of the centre 'skeg', seen exposed in the dry dock environment. Alternative arrangements are constructed with a centre line Controllable Pitch Propeller with azipods set to either side.

High Speed Craft (HSC)

Chapter 2 of the High Speed Craft Code draws attention to the potential hazards that may affect high speed design craft, when manoeuvring at speed:

1. Directional instability is often coupled to roll and pitch instability.
2. Broaching and diving in following seas, at speeds near to wave speed is applicable to most types of craft.
3. Bow diving and craft on the plane, both in mono-hulls and catamarans, is due to dynamic loss of longitudinal stability in relatively calm seas.
4. Reduced transverse stability with increased speeds in mono-hulls.
5. Pitching of craft on the plane (mono-hulls) being coupled with heave oscillations can become violent (similar to a porpoise action).

6. Chine tripping, being a phenomenom of mono-hulls on the plane occurring when the immersion of a chine generates a strong capsize moment.
7. Plough-in of air cushion vehicles either longitudinally or transversely as a result of bow or side skirt tuck, under or sudden collapse of skirt geometry, which in extreme cases could cause capsize.
8. Pitch instability of SWATH (small water plane area twin hull) craft, due to the hydrodynamic moment developed as a result of the water flow over the submerged lower hulls.
9. Reduction in the effective metacentric height (roll stiffness) of surface effect ship (SES) in high speed turns compared to that of a straight course, which can result in sudden increases of heel angle and/or coupled roll and pitch oscillations.
10. Resonant rolling of SES in beam seas, which in extreme cases could cause capsize.

Specific design features incorporated at building can go some way to overcome the above affects and enhance safer stability conditions and manoeuvring aspects.

High speed craft

HSC categories

The IMO, HSC code was introduced in 1994 and had mandatory implementation in 1996. Under the auspices of the code, High Speed Craft were placed into one of three categories:

Category 'A' craft

Defined as any high speed passenger craft, carrying not more than 450 passengers, operating on a route where it has been demonstrated to the satisfaction of the flag or port state that there is a high probability that in the event of an evacuation at any point of the route, all passengers and crew can be rescued safely with the least of:

i. time to prevent persons in survival craft from exposure causing hypothermia in the worst intended conditions;
ii. the time appropriate with respect to environmental conditions and geographical features of the route, or
iii. four hours.

Category 'B' class

Defined as any high speed passenger craft other than a Category 'A' craft, with machinery and safety systems arranged such that, in the event of damage, disabling any essential machinery and safety systems, in one compartment, the craft retains the capability to still navigate safely.

A cargo craft class

Defined as any high speed craft other than a passenger craft and which is capable of maintaining the main functions and safety systems of unaffected spaces, after damage in any one compartment on board.

Maximum speed formula

Speed must be equal to, or exceed 3.7 times the displacement corresponding to the design waterline in metres cubed, raised to the power of 0.1667 (metres per second).

Applicable to most types of craft, corresponds to a volumetric Froude number greater than 0.45.

High speed craft

The bow wave and generated wake made by a small high speed pilot craft operating in calm open waters. Such water disturbance can affect other small craft which may be in close vicinity.

A high speed passenger ferry operating off the Spanish coastline in calm, but restricted water. The dangers from the generated wake when operating at speed can be hazardous for craft being single-handedly manned by fishermen or yachtsmen.

Waterjet propulsion systems

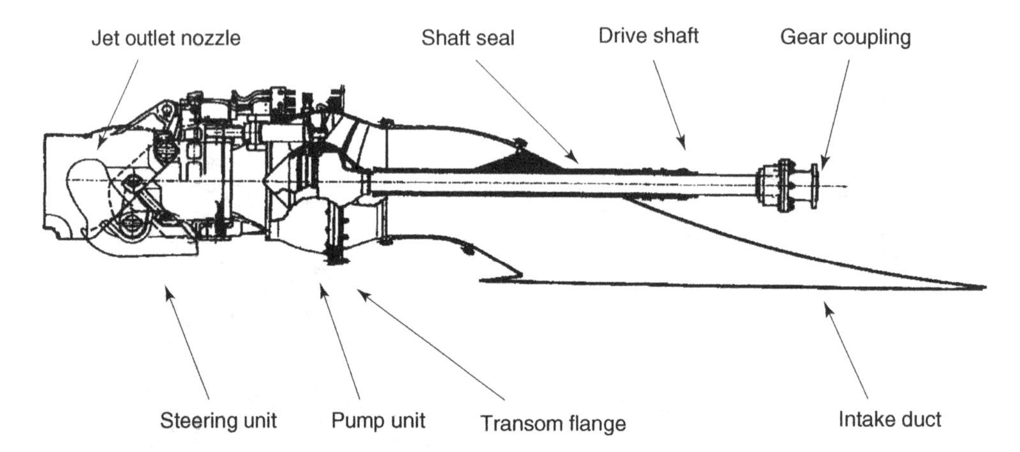

Waterjets

With the increased development in high speed craft, especially in the Ferry sector of the industry, waterjet propulsion systems have been incorporated into vessels either as a main propulsion system or alongside conventional propeller units to provide additional power and/or manoeuvring capability.

Some caution must be used with these systems when in confined waters as the jet wake generated can be powerful and could cause interaction with other traffic or coastline structures.

High Speed Craft and Safe Speed (Ref., Regulation 6, ColRegs)

It should be realized from the onset that the Collision Regulations are applicable to all vessels inclusive of HSC. This application also includes Regulation 6 'Safe Speed', which in turn must also be construed in conjunction with the other relevant remaining regulations.

The question of what constitutes a safe speed is probably irrelevant until an accident occurs. The fact remains that a high speed vessel must still retain the ability to move out of trouble just as a conventional vessel needs to avoid close quarter situations. The letter of the law within the ColRegs is designed to avoid close quarter situations and many of these can be avoided by not only a reduction of speed but also an increase of speed.

Such a statement is not meant to be controversial, but is meant to highlight that an increase of speed can be just as effective to avoid a close quarter encounter as a decrease in speed. Such action, however, should not be taken without long range radar scanning beforehand, and should not be sustained for an indefinite period. Neither should a decision of this nature be made without a full appraisal of the immediate environment.

The use of high speed in good visibility can, and is, well used to take early action to avoid close quarter situations. However, in the event of poor visibility being encountered, watch officers should to be aware of the need to be able to stop their vessel within half of the visible range, bearing in mind that a high speed craft on the 'plane' at over 40 knots, which encounters poor visibility, may reduce to say fifteen (15) knots.

High speed craft – bridge consol. A typical central position, bridge consol for a modern high speed ferry craft. Interchangeable LCD monitors for ARPA and ECDIS displays are set either side of manoeuvring controls and the centre performance and status data display.

In so doing, her mode changes to that of full displacement, and she can no longer assume the same manoeuvrability as when she is operating at increased speed.

Again this option is not being advocated by this author. On the contrary, to bring the vessel to a dead stop can, in some circumstances, be more hazardous than maintaining ship manoeuvrability. What is being highlighted is that stopping, or increasing speed, are alternative actions to decreasing speed and should not be dismissed out of hand. They are and remain, options, and the circumstances of each scenario will dictate what is considered prudent at the time.

> ***Comment***: Watch Officers are reminded, however, that Regulation Six is not a stand alone regulation, and the ColRegs also stipulate that: 'Assumptions should not be made on the basis of scanty information, especially scanty radar information'.

Wheel over points

The dangers of interaction are prevalent in many different situations, but none more so than when the vessel enters shallows and is in close proximity to the land. Tight manoeuvres must be anticipated through rivers, canals and when making land falls. The large vessel must anticipate that the position of the way point is rarely coincident with the time at which the helm will be applied. Masters would be expected to ensure that passage plans include 'wheel over points' when vessels are approaching positions of course alteration.

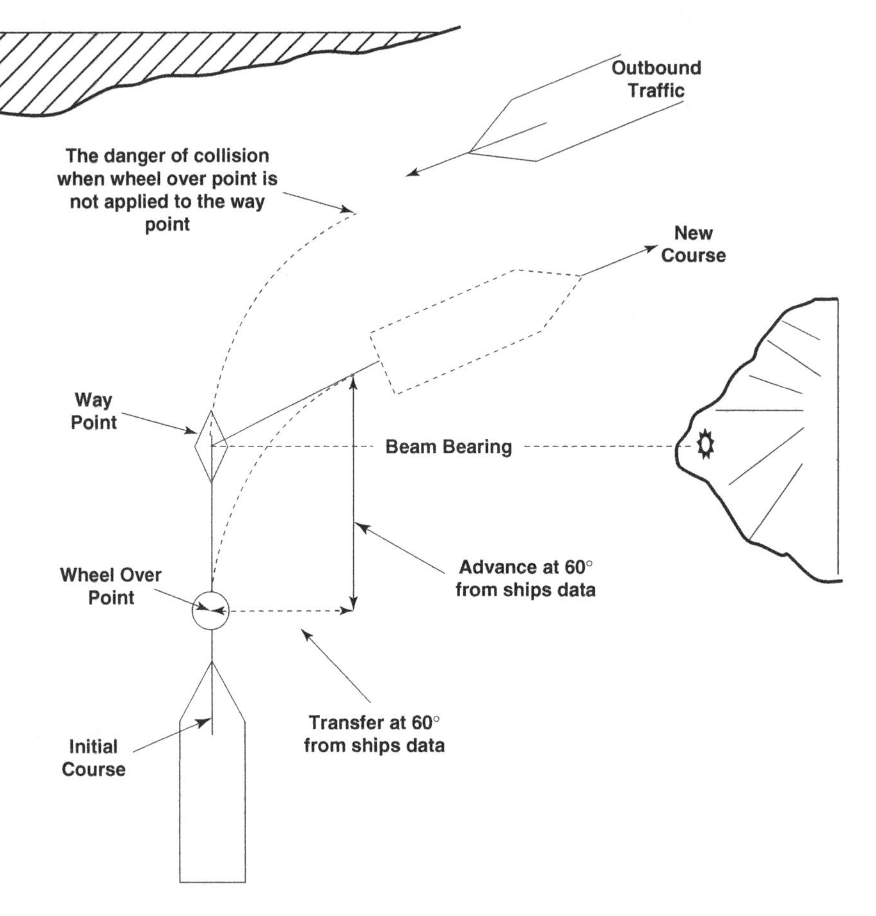

Wheel over point. This example shown for a 60° alteration of course. Advance and transfer details can be referenced from the ships sea trials and performance documentation.

Canal and river movements

Rivers and canals by their very nature have restricted water compared to open sea conditions. When the ship is in transit through a canal, the vessel occupies a volume of the canal space causing effectively a blocking restriction to water movement.

Squat and blockage

Blockage factor illustrated

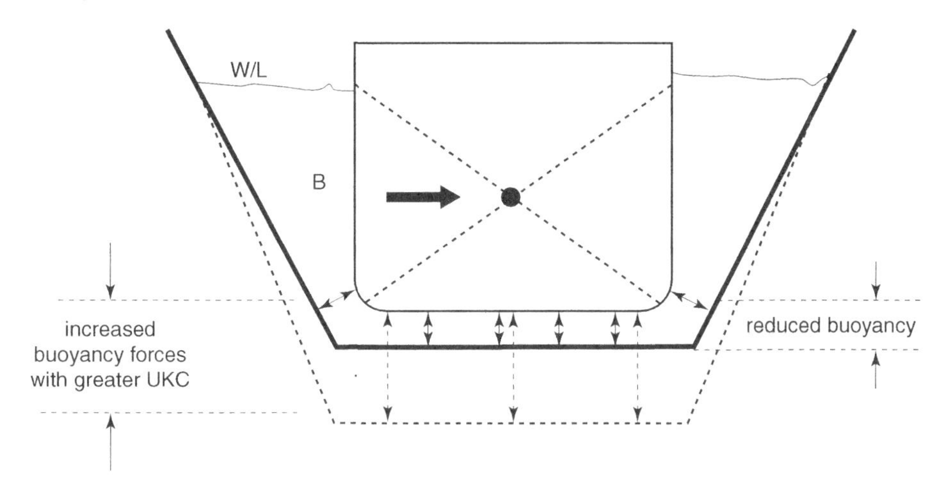

The illustration above shows that because the underkeel clearance is small, the volume of water under the keel is small and would not have the same buoyancy affect on the hull as noted in deeper water; bearing in mind that the position of buoyancy is defined as the geometric centre of the underwater volume. If the vessel is heeled by external forces the water plane will increase, the position of 'B' would move upwards but, at the same time, also outwards towards the angle of heel. This leaves the low side, at the turn of the bilge, liable to contact with the ground.

$$\text{Blockage Factor} = \frac{\text{that proportion of a midship's section}}{\text{cross sectional area of the channel, river or canal}}$$

When two ships are passing in the channel the blockage factor is increased and the value of squat experienced can expect to nearly double.

Practically the vessel may expect to encounter steerage problems caused by squat and the proximity of the canal bottom in relation to the position of the keel. Speed of movement would be critical and the vessel must expect to move at a greatly reduced speed. The use of tugs, fore and aft to effect steerage control must also be anticipated as being an absolute necessity.

The 'Stena Leader' seen departing the Roll on–Roll off river berth in Fleetwood, Lancashire. The dangers of interaction in many forms in such confined waters is ever present for the larger vessels. Reduced speed of operation is considered essential along with reliable machinery and steering gear.

Squat – ship's response

The behaviour of a ship in shallow water, where the forces of buoyancy are reduced, can expect to be totally different to the behaviour of the same ship in deeper water, where the buoyancy forces will have a much greater affect. Factors affecting the actual value of squat will vary considerably but could expect to include any or all of the following:

a) Draught/depth of water ratio. A high ratio equates to a greater rate of squat.
b) The position of the longitudinal centre of buoyancy (LCB) will determine the trimming effect and have a direct relation to the squat value.
c) High engine revolutions can expect to increase stern trim.
d) The speed of the vessel is related to the value of squat in that the value is influenced by $speed^2$. The faster the ship moves the greater the squat value.
e) The type of bow fitted effects the wave making and pressure distribution on the under water volume.
f) The length/breadth ratio can cause an increase or decrease of the squat value, i.e. short-tubby ships tend to squat more, than the longer narrow beam vessel.
g) The breadth/channel width ratio affects the squat value. A high ratio causing an increased value of squat.
h) Vessels with a large block coefficient C_b will experience greater effects from squat.
i) Greater effects of squat are experienced when a vessel is trimmed by the bow than by the stern.

Evidence of squat

The indication that a vessel is experiencing squat will show from the steering being affected. Waves from the ship's movement will probably increase in amplitude and the wake left by the vessel will probably be mud stained. Some vibration may also occur with a decrease in speed and a reduced rpm.

Interaction – situations within the marine environment

A small tug, seen engaged with the large carrier 'HUAL Trotter'. The danger of interaction is a one when small craft like tugs are engaged in close proximity to larger, parent vessels.

A large tanker manoeuvres in close proximity to an FPSO in offshore regions. Such close manoeuvres between vessels and fixed or floating structures are known to generate interactive forces which tend to hamper ship handling operations.

Where a vessel is brought into close proximity of a bank, as in a canal or river, it may experience a pressure build up between the hull and the obstructing bank, known as 'bank cushion effect'. This pressure build up would effectively turn the bows of the vessel away from the bank and force the ship's heading into the middle channel area, away from the restrictions of the channel sides.

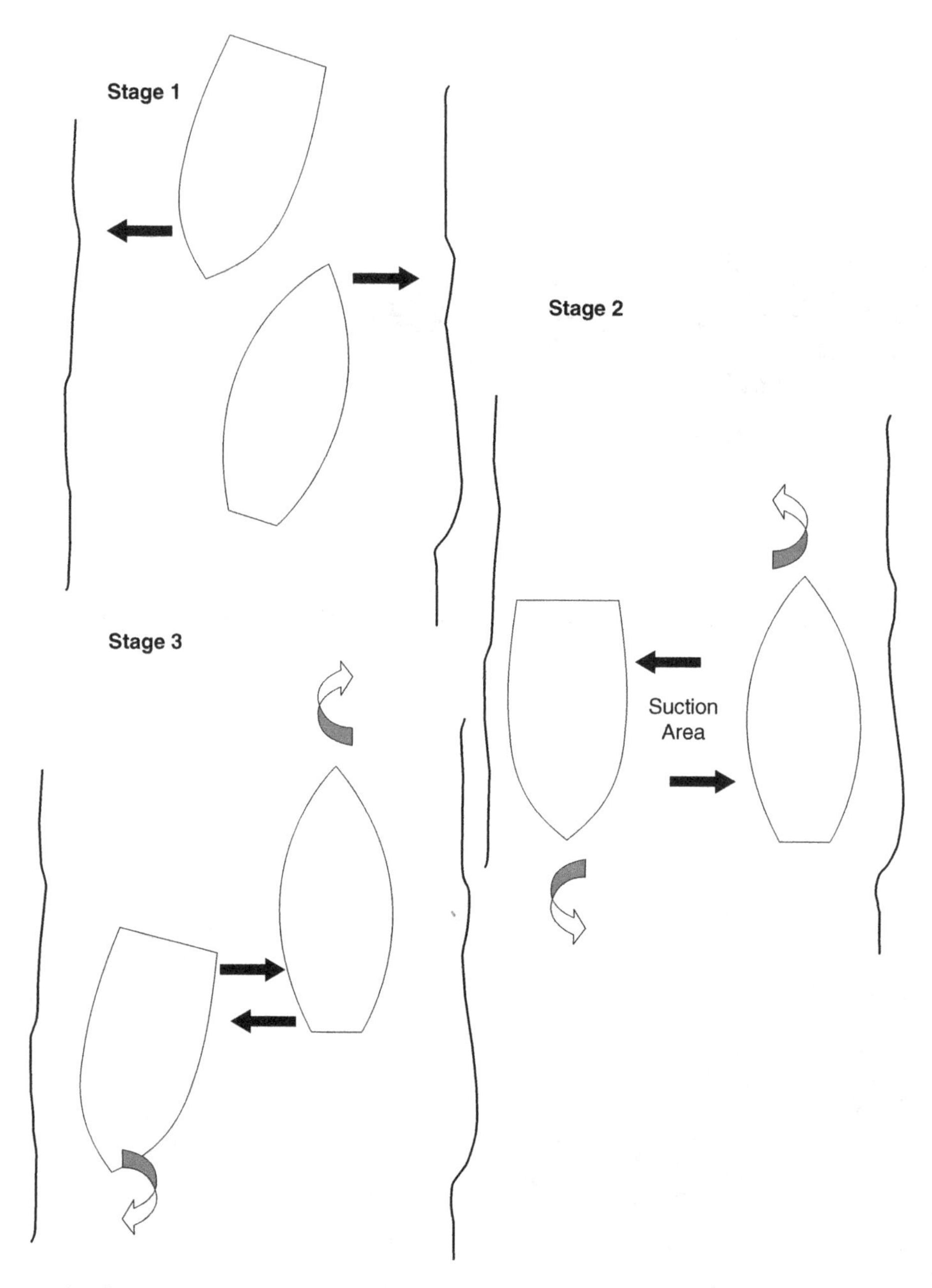

Interaction forces – vessels meeting 'End On' passing too close.

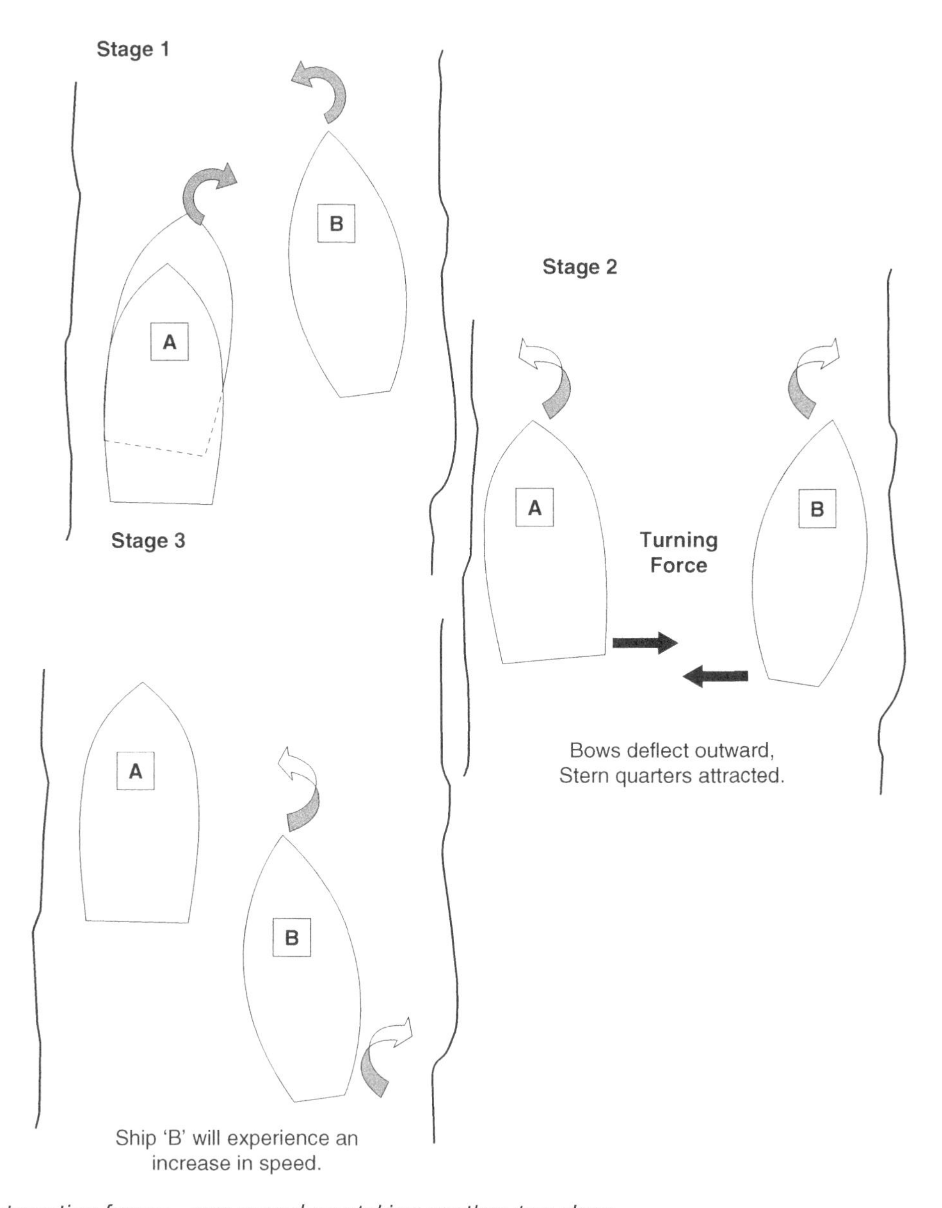

Interaction forces – one vessel overtaking another, too close.

In itself, this can be countered by applying helm, if it is expected and catered for. However, the movement from a bank cushion effect could have serious consequences if, say, the vessel is being overtaken or meeting an oncoming vessel moving in the opposite direction, the vessel close to the bank taking a sheer towards the oncoming traffic.

The pressure cushion generated cannot be avoided, but the violent reactive movement can be curtailed by reducing the speed of approach towards the bank. The speed of the vessel being reduced to steerage way only, will expect to minimize the outward turning effect of the vessel.

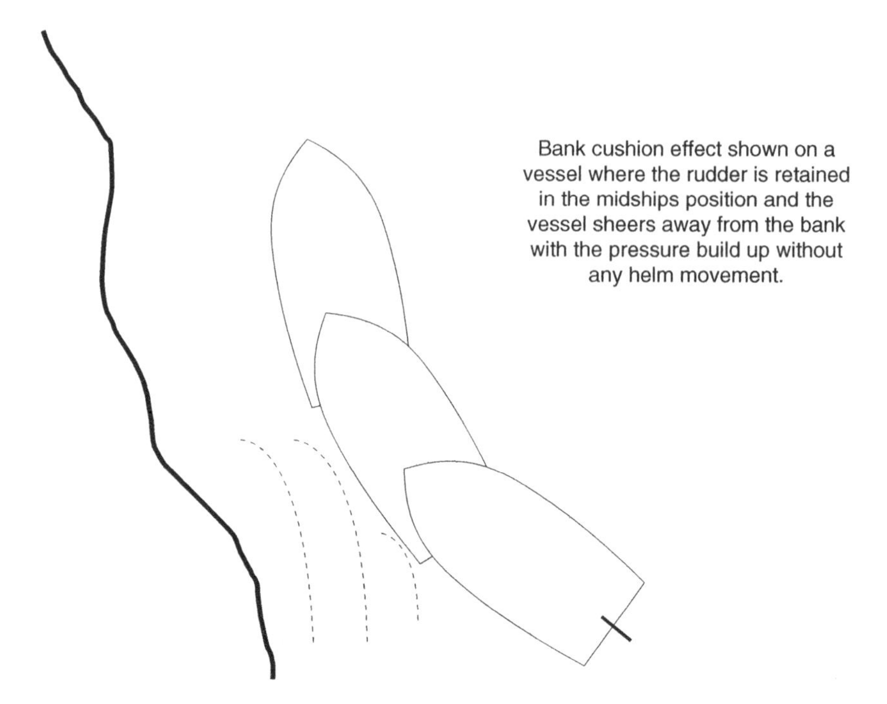

Interaction forces – bank cushion effect.

Shallow water effect

When ships make a landfall from a deep sea position they may experience a form of interaction with the sea bed, known as 'Shallow Water Effect'. It is especially noticeable where the shoals and the change in depth becomes abrupt and may cause the ship's steering to be affected, the bows being pushed off course to either port or starboard as the vessel experiences a sharp change in underkeel clearance.

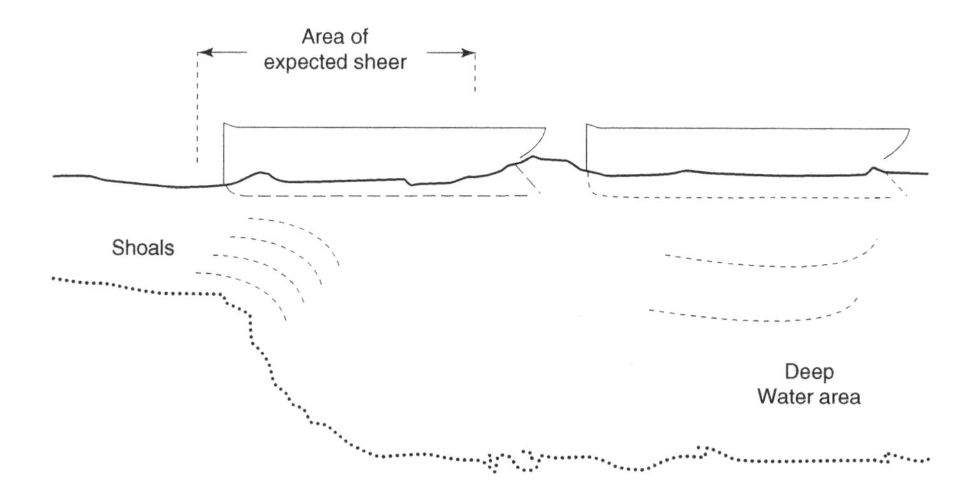

As the vessel approaches the shoal area, the interaction between the hull and the closeness of the sea bed may cause the vessel to sheer away. A reaction that can be quickly corrected by alert watchkeepers but could generate a close quarters situation if other traffic is in the near vicinity.

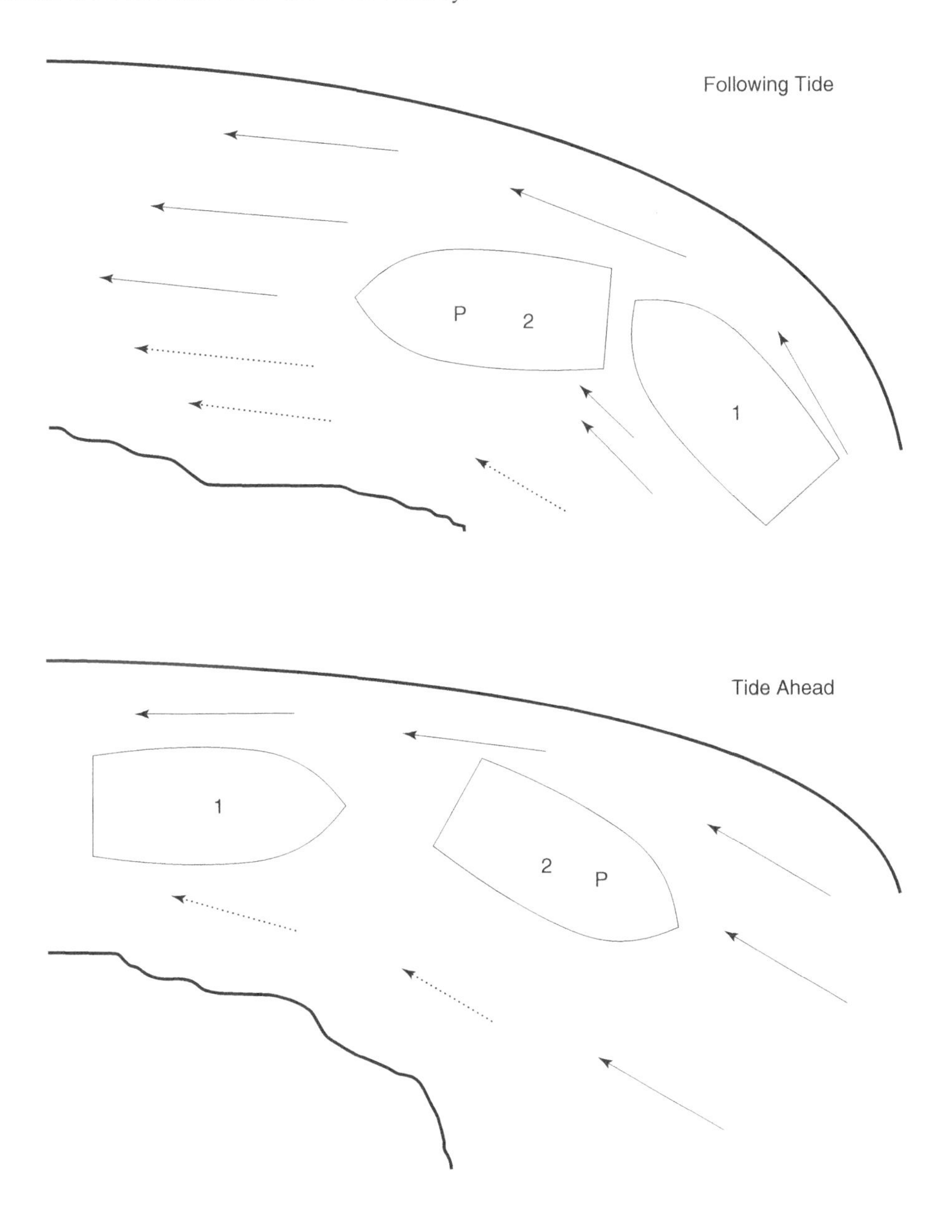

Negotiating bends in tidal riders. 'P' represents the position of the ship's pivot point when going ahead.

Interactive forces between tug and parent vessel

With the following example a tug is to engage with a parent vessel on the starboard bow. Interaction between the smaller and larger vessels could generate a collision scenario. Prudent use of the helm and speed by the tugmaster will be crucial in collision avoidance.

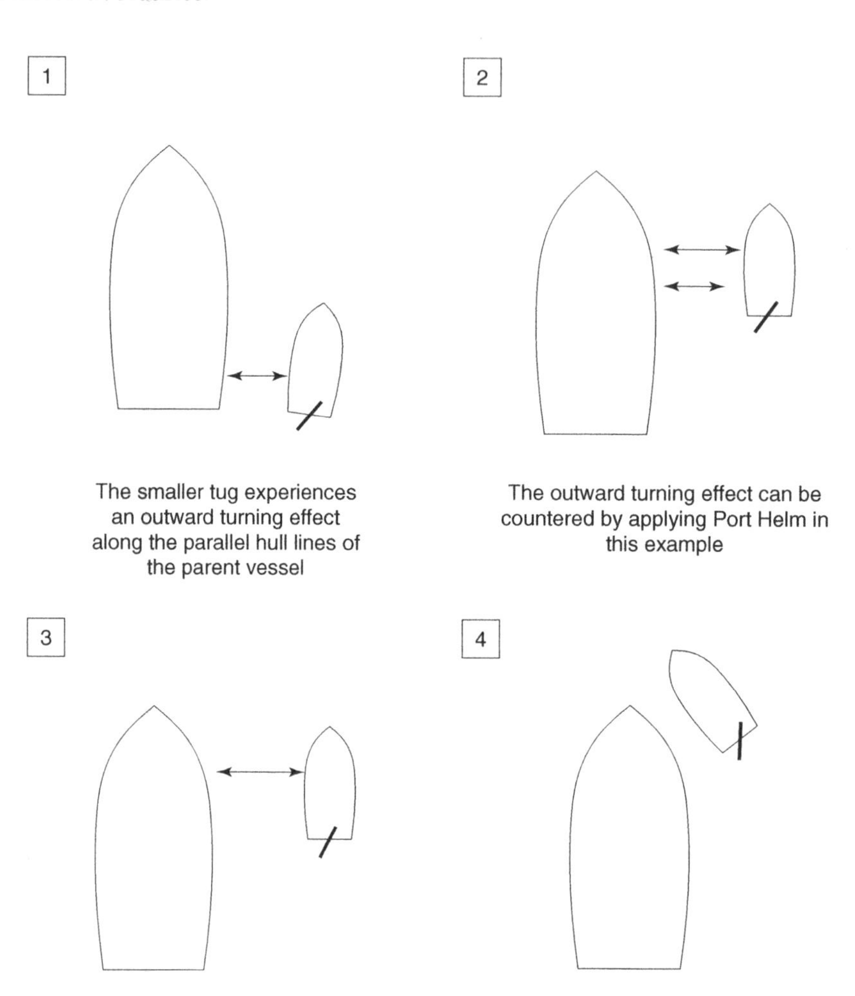

1 The smaller tug experiences an outward turning effect along the parallel hull lines of the parent vessel

2 The outward turning effect can be countered by applying Port Helm in this example

3 The tug experiences the maximum outward turning force at the shoulder position of the larger vessel

4 Because the tug is carrying Port Helm and still moving ahead, a loss of turning force is experienced under the flare of the ships bow. The tug could sheer across the bows of the larger vessel with inevitable collision

3 Anchor operations and deployment

Introduction; Anchor terminology; Anchor types – securing and stowage arrangements; Anchor planning – single anchor procedures, safety precautions when employing anchors; Anchor watch – dragging anchors; Anchor work operations; Baltic and Mediterranean moorings, running and standing moors; Dredging an anchor; Large vessels; Kedge anchor; Emergency use of anchors; Mooring operations; Hanging off an anchor; Clearing foul hawse; Clearing fouled anchor; Mooring by cable to buoys; Use of slip wires.

Introduction

It has been suggested that the first anchors might have appeared as a basket of stones used by vessels engaged on the River Nile around 6,000 BC. In the twenty-first century, the deployment and holding power of anchors has improved somewhat from the basket of stones of the ancient Egyptians.

The high holding power anchors, the use of multiple anchor moorings and the sheer size of anchors for the larger vessel, have all brought with them associated operations and relevant complications. Anchor cables are brought into use, sometimes without the anchor as in mooring to buoys, while the problems of fouled anchors, foul hawse and lost anchors present concerns for seafarers as well as insurers.

Developments in the offshore industry have spilled over into aspects of commercial shipping, while the early experiences of shipping fuelled advances in the roots of offshore. Historically, terminology from the sailing ship days has lingered on, but technological advances in windlasses, braking systems, anchor design and the need for greater holding power have all changed the face of modern day, anchor operations (see Appendix C).

A ship's use of anchors when berthing has become common practice. More ships will expect to moor with two anchors as opposed to coming to a single anchor which requires a larger swinging room. Vessels will still experience the dragging anchor, when the elements affect the exposed vessel; we cannot control the weather but we can be prepared for the worst and the anchored ship only remains safe when the personnel within continue their effective duties as watchkeepers.

Even though anchor plans have become more formalized by ships' Masters, the weather and tidal affects cannot in any way be guaranteed. Watchkeepers must be effective in their duties, especially in the task of monitoring of the ship's position. Far better to be party to good seamanship than have to rectify mistakes of inefficiency in the field of anchor use. If and when problems arise, the remedy is often heavy and labour intensive; anchor cables are heavy and difficult to manipulate.

Anchor work – terminology associated with ship handling operations

A'cockabill	The term employed to describe the anchor in a cleared position, prior to letting go.
Anchor	A heavy object designed to prevent a ship or structure from drifting from a desired position. This objective is achieved by lowering the anchor to the sea bed by a length of chain cable or warp. The design of anchors differs to allow, usually, a spade or hook effect into the sea bed.
Anchorage	A geographic area suitable for ships to lay at anchor. Ideally, it would have good holding ground free of strong currents and sheltered from prevailing weather. It is usually identified on a navigational chart by a small blue anchor symbol.
Anchor and Chain Certificate	A Certificate issued by a Classification Society or Government Maritime Authority which reflects that the anchors and cables listed have passed tests and inspections satisfactorily.
Anchor and Chain Clause	A marine insurance policy clause which may exempt underwriters from the costs of recovering anchors and cables if lost by the vessel when afloat.
Anchor Aweigh	The anchor is said to be 'A-Weigh' at that moment when it is broken out of the holding ground and hangs clear of the sea bottom.
Anchor Ball	A black ball shape shown as a day signal by a vessel at anchor. The diameter of the shape is not less than 0.6 m and it is exhibited where it can best be seen in the fore part of the vessel.
Anchor Bearings	A set of crossed compass bearings as observed from a vessel laying at anchor to identify the ships position.
Anchor Bed	A sloping platform used to accommodate the stowage of the Anchor. An old fashioned term which tended to be dropped from general use with hawse pipe design being widely adopted (older vessels sometimes had a 'Tumbler' fitted to the inboard side of the anchor bed. Namely a revolving bar fitted with horns to hold anchor securing chains).
Anchor Bell(s)	A bell signal used to indicate the number of shackles of cable extending from the ship to the anchor. The ship's bell, situated on the fo'castle head, was generally used for this communication to the bridge personnel. It is still used on the smaller ships, but portable radios have replaced bell signals, especially on the larger type vessels.
Anchor Buoy	A small buoy used to indicate the position of the anchor when on the sea bottom. The length of the buoy line being adjusted to the depth of water, and long enough to compensate for any rise and fall in the tide. This is ideally a marker buoy, and should not be confused with an anchor recovery buoy as employed in the offshore industry.
Anchor Coming Home	A descriptive term to indicate that the anchor is being drawn home towards the ship in the operation of heaving aweigh. The action is unusual as it is normal to expect the ship to be drawn towards the anchor.

Anchor Dragging	A term describing a vessel moving her position, because the anchor is no longer secure in the sea bed and holding the vessel. The vessel is dragging her anchor as her position moves.
Anchor Dues	Fees paid by a ship for anchoring inside a port or harbour Limits.
Anchor Light(s)	That name given to the riding light(s) when a vessel rides to her anchor during the hours of darkness between sunset and sunrise.
Anchor Parts (Conventional Anchors)	Anchor 'crown' shackle Arm (s) Crown Fluke Forelock (for securing a stock) Gravity Band Head (comprising of the lower parts, excluding shank) Palm (Admiralty Pattern/Fisherman's type anchors only) Pea or Bill Shank Stock Trend (or Throat) Tripping palms.
Anchor Party	A group of crew members, usually supervised by a Deck Officer, who are employed to operate anchors and cables when letting go or heaving anchors.
Anchor Pendant	An anchor towing/lifting wire which is secured to the Crown of an anchor. The other end of the wire is then craned to an anchor handling vessel in order to heave the anchor clear of an offshore rig prior to laying the anchor.
Anchor Plan	A proposed plan usually constructed by the ship's Master or navigator, to bring the vessel into an anchorage position safely.
Anchor Plate	A metal plate at the side of a vessel where the inboard fluke of a stocked anchor would rest. *See: Anchor Shoe.*
Anchor Pocket	A bow recess designed to accept the head of the anchor.
Anchor Recess	Sometimes referred to as the 'Anchor Pocket'. That space at the lower end of the hawse pipe designed to accept the stowage of a conventional stockless anchor.
Anchor Shackle 'D'	A heavy duty 'D' shackle secured to the top of the anchor shank. Sometimes referred to as the 'Anchor Crown "D" Shackle', securing the chain cable to the anchor. This is a common fitment which is supplied with most commercial anchors (smaller anchors may be alternatively fitted with a ring).
Anchor Shoe	Steel doubling plates secured to the outer hull of the vessel to prevent the 'Pea or Bill' of the anchor flukes causing damage to the ship when heaving the anchor back on board.
Anchor Spread	A pattern of several anchors employed to hold an offshore installation in position. The anchors being laid by anchor handling vessels and the number would depend on several factors, namely: depth of water, size of installation, holding ground, current/tide movements, weather conditions, period of operation. The spread would relate to the scope of cables and the angular distance apart from the installation.
Anchor Types	Admiralty Class (Cast) anchor (A.C. 14) (A.C. 16A) (A.C. 17) Admiralty mooring anchor (e.g. type A.M.12 = 6 tonnes) Admiralty Pattern anchor (Fisherman's anchor) (Alt. Common anchor, Admiralty Plan)

(Continued)

Table (Continued)

	Baldt anchor Bow anchor Bruce anchor (FFTS) (FFTS Mk 4) Byers anchor Close Stowing anchor Clump anchor (concrete) C.Q.R. (USA – plough anchor) Creep anchor Danforth anchor (close stowing stocked anchor) Denla (Bruce) – drag embedment anchor D'Hone anchor Eells anchor Flipper 'Delta' anchor Gruson anchor Hall anchor Heavy cast 'Clump' anchor (iron) Herreshoff anchor (moveable stock) Heuss Special anchor Improved mooring anchor Jury anchor (improvised emergency anchor) Kedge anchor (Stern) (smallest anchor on board) Klipp-Anchor Meon Mark 3 anchor Mushroom anchor Northill anchor (seaplanes) Piggy anchor (alt. back anchor) Pool – N anchor Pool – TW anchor Sheet anchor (spare) (obsol., best anchor) Single fluke anchor Spek anchor Stato anchor Stem anchor Stevin Stocked close stowing anchor Stockless anchor (double fluked anchor) Stokes anchor Stream anchor (*alt.*, Stern anchor) Tombstone anchor Trotman anchor Umsteckdraggen (4 arm/fluke anchor) Union anchor
Anchor Warp	A rope or hawser used as an alternative to anchor chain cable.
Anchor Watch	A term used to describe the period of time that persons would look after the safe keeping of the vessel, when at anchor. The anchor watch is usually made up of several crew members including an 'Officer of the Anchor Watch'.
A-peak	A term which describes the position of the ship when the vessel's bows are above the anchor position during the operation of weighing anchor. The cable would be 'up and down,' just prior to the anchor breaking out.
A-Trip	That moment that the anchor breaks out by the heaving of the cable when weighing anchor.

Baldt Shackle	A sectioned anchor/cable joining shackle.
Baltic Moor	A combination mooring of a vessel alongside a berth which employs a stern mooring wire shackled to the offshore anchor cable in the region of the 'ganger length'. When berthing the offshore anchor is let go and the weight on the cable and the stern wire act to hold the vessel just off the berth.
Band Brake	A braking system found on a windlass or cable holder which is applied to the 'cable drum' to slow and check the run of cable, when letting go the anchor.
Bite	An anchor is said to 'bite' when it digs into the sea bottom and holds position.
Bitter End	That end of the anchor cable which is secured in the 'clench' found at the chain locker.
Blake Slip	A cable holding slip, often called a 'Pelican Hook' in the US. It is sometimes used in conjunction with a bottle screw (US turnbuckle) to provide additional securing to a stowed anchor or hold an anchor prior to release. It can act as a weight preventer on chain cable but it is only tested to half of the proof load of the cable (mainly employed by warships).
Bolster Bar	A stowage bar for anchors, found in a position clear of the water surface on the leg columns of offshore oil/gas rigs and around the bow of offshore supply vessels and cable ships
Bonnet	A type of hooded cover over the deck opening of the 'hawse pipe' or 'navel pipes'. Restricts the ingress of deck water into the chain locker. These are mostly fitted to warships and must be considered as old fashioned.
Bow Stopper	Collective term to describe an anchor cable holding/securing device. Different types of 'bow stoppers' are found but the most common are probably the: guillotine, compressor, or tongue types (guillotine stoppers sometimes called 'Dog' type).
Brought Up	That term which describes the vessel is anchored to a desired scope of cable and holding position, i.e. not dragging anchor.
Bruce Anchor	A high holding power anchor used extensively in the offshore industry.
Bullring	Centre lead at the forward end of the vessel. Generally used for running additional mooring lines but may be employed for towing or for anchor cable when mooring to a buoy (US bullnose).
Cable Clench	The holding arrangement for the 'bitter end' of an Anchor cable. Usually found inside or on the outer bulkhead of the chain locker. It operates with a quick release system.
Cable Holder	Alternative winch system to a windlass employed for heaving and letting go the anchor. Cable holders often have a capstan combined in the design, for additional use of mooring lines. Cable holders are more common to warships or the large passenger liners.
Cable Jack	A tool employed for lifting 'jacking up' heavy anchor cable. Usually found around 'Dry Docks'.
Cable Lifter	*See: Gypsy.*
Cable Meter	A device for measuring the amount of cable paid out. Employed mainly with automatic anchoring operations.
Cable Stopper	US term for 'Bow Stopper' alt., Chain Stopper.

(Continued)

Table (Continued)

Cable Tier	A below deck stowage position for anchor warp found on the old sailing ships. The term is still used by mariners but to express the Chain Cable Locker.
Capstan	A vertical mooring drum driven by either hydraulic or electric power (early versions on sailing ships used 'Capstan Bars', to which manual labour was employed to turn the drum/barrel). The drum is scored with 'whelps' to provide increased grip and holding ability on the hawser being worked. Capstans are often designed as a combined dual element of a 'cable holder' and 'warping drum'.
Carpenter Stopper	A heavy wire hawser, stopper arrangement. It operates on a wedge principle and clamps the wire in position. It is widely used in the salvage industry where anchors and warps are being employed as ground tackle.
Cathead (Clump Cathead)	An obsolete practice of 'catting' an old fashioned stock anchor to a 'Cathead' fitted to the starboard bow area of a sailing ship. The operation has been superseded by 'Stockless' anchors and 'hawse pipe' arrangements. 'Catting' is defined as hoisting aboard.
Chain Hook	Long handle steel hook used to manhandle chain cable.
Chain Locker	The stowage space used for the ships anchor cables.
Chain Roller	A deck fitment (not always used) which if fitted is situated between the bow stopper and the hawse pipe, to facilitate the paying out of the anchor cable.
Chaser	A ring which has attached both the work wire of the anchor handling vessel and also the anchor pendant. Once the anchor is laid the handling vessel chases the pendant back to the rig by pulling the ring along the anchor warp length of the set anchor (specific to offshore activity).
Clenching	The term given to the heating and burning over of the protruding end of a shackle bolt. It takes the place of a forelock or retaining pin in anchor or joining shackles. It is also a practice employed to secure buoy mooring rings to chain legs.
Common Link	A link of the anchor chain cable, maybe studded or open link chain.
Creep Anchor	A specialized recovery anchor like an enlarged 'grapple'. It is a four pronged anchor used by buoy laying/recovery vessels for service and maintenance of buoy moorings (specific to Cable Ship operations).
Crown of an Anchor	That area of the anchor head found at the base of the shank between the tripping palms.
Devils Claw	A holding claw which secures the anchor cable and provides additional securing to the anchor when the vessel is at sea.
Die Lock Chain	A modern day type of construction of anchor cable achieved by links being manufactured in two sections and forged together.
'D' Lugged Joining Shackle	A cable joining shackle, also employed to shackle the cable to the anchor. It is an alternative to the Kenter shackle, but must be used with open links either side when joining cable lengths.
Dovetail Chamber	A shaped recess at the top of Joining Shackles which contains the lead pellet which in turn prevents the spile pin from accidentally dropping out.
Dredging an Anchor	The deliberate use of the ship's anchor on the sea bed, deployed at short stay, to influence the movement of the ship's head. Dredging being the act of moving both the vessel and the anchor over the ground.

Drift	A steel extension piece which is employed for (i) expelling the spile pin from a joining shackle (ii) inserting lead pellets into joining shackles.
Drop an Anchor Underfoot	To let an anchor go without veering cable. It is often used as a second anchor to reduce the vessel's 'yawing' movement when lying to a single anchor. This use of a second anchor would be usually held at 'short stay'.
Dunes Anchor	Trade name for a stockless anchor.
Ebb Tide	When the tidal flow is out of an estuary or harbour, away from the land (*see also Flood Tide*).
Eductor	Pump arrangement based on the venturi effect used to pump out the mud box of a chain locker on a deep draughted vessel.
Eell's Anchor	A patent stockless anchor with long flukes used extensively in salvage work.
End Link	An enlarged open link found at the bitter end of the cable which is held by the 'cable clench'.
Fairleader	A roller lead found on the side of an 'Offshore Platform' designed to provide a lead for the anchor warp to the mooring anchor.
Flood Tide	When the direction of tidal flow is inward into an estuary or harbour (tidal flow towards the land).
Foul Anchor	When the anchor is found to be obstructed or entangled with debris or other foreign body dragged from the sea bed, when weighing anchor.
Foul Hawse	The description given to when the two anchor cables have become turned and twisted together with both anchors deployed. Cross – descriptive term to indicate cables have crossed by the vessel swinging through 180° Elbow – descriptive term to indicate cables are fouled by the vessel swinging through 360° Cross and Elbow – descriptive term to describe a foul hawse where the vessel has swung round through 540° Round turn – descriptive term to indicate a foul hawse where the vessel has swung through 720°.
Ganger Length	A short length of cable found between the anchor crown 'D' shackle and the first joining shackle of the cable.
Gravity Band	An iron or steel ring passing around the shank of an anchor at that point where all the forces of gravity act and the anchor can be suspended in a balanced position. Found more often on the old Admiralty Pattern Anchors (Fisherman's anchor).
Ground tackle	Extensively employed in salvage operations to hold a vessel's position on station. Ships anchors with lengthy cable can be used for this purpose but generally, additional anchors and cables are brought in, specifically for the task.
Grow	A term which describes the exposed amount of cable, in the direction of the anchor. It is seen to extend/grow as the anchor digs in and starts to hold.
Gypsy	A common term used by the seafarer to describe the geared wheel while encompassed with the links of the anchor cables when heaving in or paying out on the cable. Fitted to cable holders and windlasses (alternative term is Cable lifter).
Hartford Shackle	A buoy mooring shackle.

(Continued)

Table (Continued)

Hawse Hole	Forerunner to the 'Hawse Pipe' which was employed to lead anchor warp outward towards the anchor on sailing ships.
Hawse Pipe	An anchor stowage pipe usually set into each bow of the ship to accommodate the stowage of the stockless anchor.
Hove in sight	That moment in time when weighing anchor, that the anchor is sighted by the Officer in Charge of the anchor party (*see Sighted and Clear*).
Joggle Shackle	A curved shaped shackle employed in cable operations.
Kedging	Moving of the ship by means of small anchors and hawsers. A vessel may attempt to 'kedge' stern first off a sand bar after running aground.
Kenter Lugless Joining Shackle	A common joining shackle when joining shackle lengths of anchor cable.
Killick Anchor	A Navy term for a light anchor that can be carried out by a small boat for the purpose of 'Kedging' or use as an extra mooring.
Lead Pellet	A small quantity of lead used to prevent 'spile pins' from dislodging accidentally inside joining shackles.
Load Cell	A cable tension, measuring device used extensively in the offshore industry when laying anchor patterns to provide indication whether an anchor is dragging or holding.
Long Stay	When the vessel rides to anchor where the line of cable towards the anchor, lies nearly parallel to the water surface.
Maul	A hammer-like instrument which can be used to punch out and drift spile pins from joining shackles (Wt 6–7 lbs/ 2.72–3.17 kg). Shipwrights pin maul had a single flat face with a long drawn out pin opposite. This posed a danger to other persons in the vicinity when in use and it was later changed to a double face.
Mediterranean Moor	A stern to quay mooring for a vessel. It employs two bow anchors, one from each bow at about 25–30° from the fore & aft line. The stern of the ship is manoeuvred close to the quayside and secured by stern lines. It is popular with Ro–Ro vessels for stern ramp operation and for tankers with stern discharge facilities.
Mooring	(i) A term to describe a vessel which is anchored with more than one anchor. (ii) A term used to describe the operation of tying the vessel up alongside a berth. (iii) 'Mooring Deck' descriptive term used to describe the deck area where anchors and cables are worked or where mooring ropes are employed to secure the ship.
Mooring Swivel	A four chain leg with centre swivel arrangement employed for mooring a vessel by use of its two anchor cables and two mooring buoys. The piece is fitted with union plates at each end to facilitate the triple connection between chain legs and swivel.
Navel Pipes (RN)	A deck pipe which carries the anchor cable down into the cable locker. Mercantile Marine use the term 'Spurling Pipe' or 'deck pipe' to mean the same.
Open Moor	A type of ship mooring with an anchor paid out from each bow. Each anchor is set at about 25° off the fore & aft line. This moor is useful in non-tidal waters, e.g. Fresh water river.
Palm	Inside face of the fluke of a stocked anchor.

Piggyback	A term which describes the use of a second anchor in 'tandem' to prevent the first anchor from breaking out accidentally. Sometimes referred to as a 'Back anchor' or more commonly as a 'Piggy anchor'.
Pitch	The descriptive name given to a ship in a seaway when the bow area is observed to move violently in the vertical, due to the direction and height of sea conditions. Pitch can also occur when a vessel lies at anchor and the action is to cause a heaving force on the cable and anchor.
Pointing Ship	Providing an angle to the ship away from the weather by means of running a stern mooring wire to the anchor cable, then paying out more cable. A useful operation if launching boats and wanting to create a lee when at anchor.
Range Cable	An act of paying out and flaking the anchor cable. This is usually carried out when a vessel is in Dry Dock to facilitate inspection of anchors and cables.
Reaction Anchors	Counter weight anchors to resist vessel turning forces used with salvage services if righting or turning a vessel.
Reaming tool	A hand implement used to ream out residual lead from the 'dovetail' chamber of the Kenter Joining Shackle, after the spile pin/lead pellet have been expelled.
Render Cable	Light use of the brake of a windlass/cable holder to allow the cable to pay out under its own weight.
Riding Bitts	Employed with anchor warp to turn up the warp and hold the ship once the anchor has been let go. Used on large square rigged early sailing ships 1750s, it is now obsolete.
Riding Cable	The weight bearing cable when a vessel is moored by two anchors as in a Running Moor, or Standing Moor, operation. *See: Sleeping Cable.*
Riding Slip	A cable slip generally employed on warships to hold control of the anchor cable when in the chain locker.
Riser	A general term given to a wire or chain cable rising from the sea bed moorings, to a surface buoy.
Scope	A term used to describe the amount of cable which has been paid out from the entrance of the 'Hawse Pipe' to the anchor crown 'D' shackle. Defined by the ratio of chain length to the depth of water.
Scotchman	A steel sheathing used to protect wooden decks from excessive wear when the anchor cable is running out. Usually found with cable holder arrangements rather than windlass operated anchors.
Sea Anchor	A drogue used in any craft from a liferaft to an ocean-going vessel. Smaller vessels will employ a fabric bag made of canvas or similar material to act as a drag, to retain the boats head to wind and sea intended to reduce drift. 'Jury Sea Anchor' an emergency course of action which may employ anything, e.g. a coil of mooring rope, to reduce the drift.
Shackle of Cable	A length of anchor cable. The number of shackle lengths usually shackled to the anchor will vary from ship to ship. The average vessel will carry about ten (10) shackles on each anchor. A larger vessel like a VLCC, may have up to eighteen (18) shackles. Shackle length = (15 fathoms) or (90 feet) or (27.5 metres).

(Continued)

Table (Continued)

Sheer	A term which can be applied to a vessel at anchor. It describes an angular movement by the vessel about the position of the bows which can be deliberately caused by applying helm to port or starboard.
Sheet Anchor	Obsolete term for an additional, second anchor carried by larger vessels as a back up, and kept ready for use. Unlike the modern day vessel which carries a 'spare' anchor, but not usually kept operational.
Shorten Cable	To heave in on the anchor cable and reduce the scope.
Short Stay	The cable is described to be at short stay when it is hove in close to the ship. Close to, but not quite, 'up and down'.
Sighted and Clear	That moment when weighing anchor that the anchor clears the water surface and is seen by the 'Anchor Party Officer' to be clear of obstructions and not fouled in any way.
Single Anchor	The operation of bringing a vessel into a single anchor where she anchors by means of only one anchor.
Sleeping Cable	The term given to the second anchor cable when she is moored by two anchors, i.e. Running Moor, Standing Moor. Only one of the two anchors set will be weight bearing, the second with no weight is known as the 'sleeping cable'.
Snub	The action of stopping the cable running out by applying the brake. To 'snub round' on the anchor is to check the forward movement of the vessel as the brake holds the anchor cable and prevents further cable being paid out.
Snug	The recess space on the gypsy, or cable holder which chain links lock into when the anchor is being heaved inboard.
Spile pin	A tapered pin manufactured in mild steel which binds the stud and the two halves of the Kenter joining shackle when assembled. Also used with the 'D' lugged joining shackle.
Spring Buoy	An intermediate anchor buoy employed between the anchor and the mooring buoy (usually the spring buoy is in a position unseen, beneath the water surface).
Spurling pipe	Mercantile Marine term for 'Navel Pipe' or deck pipe, which is a steel pipe which carries the anchor cable down into the 'Cable Locker'.
Stopper Plug	A short length of chain employed when tanker vessels are mooring up to a Floating Storage Unit (FSU). The Tanker first hauls a messenger on board, then the chain stopper is heaved in and secured to a Chain Hawse Stopper Unit.
Stud	The centre piece of a link of studded cable.
Surge	(i) A term which can be applied to mooring ropes and anchor cable to allow the cable or hawser to run out under its own weight. (ii) The term given to the horizontal movement of the vessel at the surface, away and towards the direction of the anchor.
Swinging Room	That area that a vessel can turn around on her anchor without fouling obstruction or grounding.
Swivel piece	There are several different designs of swivels in anchor work operations. Most will have common or open links either side of the swivel and they are joined into the extremities of the cable. The swivel piece is usually set between the anchor crown 'D' shackle by a lugged joining shackle and the lugless joining

	shackle of the first length of cable. A swivel piece will also often be found at the 'bitter end' where the cable is secured to the cable clench. NB. Not to be confused with 'Mooring Swivel'.
Tide Rode	A vessel is said to be 'tide rode' when riding to her anchor and laying head into the direction of tide.
Top Swage	An implement which is employed when breaking joining shackles. It is shaped to the curvature of the shackle's surface and acts as a divider between the hammer and the shackle to prevent damage to the surface of the shackle.
Trend	The diametric difference between the upper and lower diameters of the shank, measured in millimetres to indicate the taper of the shank construction. The numerical value of the 'trend' is itemized on the Anchor Certificate.
Up and Down	A term which describes the angle of cable just before the anchor is aweigh.
Veer Cable	A method of paying out the anchor cable or a hawser under controlled power.
Walk Back	A term which describes turning the windlass in reverse to walk the anchor back, clear of the 'Hawse Pipe', under power.
Wardill Stopper	Descriptive term for a compressor type, bow stopper, taking its name from its designer.
Warping	Moving the vessel by means of hawsers and power winches.
Weighing Anchor	A term which describes heaving up the anchor and stowing it back aboard the ship.
Whelps	Projections on the sides of capstans or warping drums to bite into the surface of a mooring rope and provide an improved pulling weight on the mooring.
Windlass	A deck mounted machine operated by electric, hydraulics, steam or pneumatic power, for handling anchors and cables and mooring equipment of the vessel.
Wind rode	A vessel is described as being 'wind rode' when she is riding to her anchor, head to wind.
'Y'	International code flag, single letter meaning: 'I am dragging my anchor'.
Yawing	A vessel may 'yaw' about her anchor. It is a term which describes the movement of both ship and cable moving from side to side in an arc about the anchor position. Not to be confused with 'sheer' where the vessel moves from side to side about that point of the hawse pipe.

Anchor types

The marine industry employs many types of anchors in a variety of forms (see Terminology). However, the common factor with all anchors is their respective holding power. Historically, anchors have developed through the centuries from the basket of stones of the ancient world's first ships, through to the hook effect of the 'Admiralty Pattern Anchor' and on to the current widely used Stockless anchors.

The massive expansion in offshore environments has probably been the greatest incentive to anchor modernization. The varied types of 'Bruce Anchor', the Flipper Delta anchors and the many mooring type anchors in use, has reflected major development in the mooring of modern day ships.

Admiralty pattern anchors – Still used in the smaller coastal craft and fishing industry. It is sometimes referred to as a 'Fisherman's Anchor'. It is fitted with a stock, which is forelocked at right angles to the arms, causing one of the two flukes to 'dig in' to the ground. The remaining exposed arm and fluke are non-effective and could cause the anchor warp or cable to become fouled about itself.

Stockless anchors – Many types of stockless anchors have been developed over the years and all have respective peculiarities depending on manufacture. The holding power is about four times its own weight and as such, it is not considered a high holding power anchor. The distinct advantage is that they are readily stowed in the bows of the vessel in hawse pipes and as such kept easily available for immediate use.

High holding power anchors – (A.C. 14, Bruce) Designs of high holding power anchors vary but they average about ten times their own weight, and are considered essential for the larger vessel, e.g. Supertankers, large passenger vessels, aircraft carriers, etc.

If compared with the more common stockless anchor which is usually manufactured as a solid casting, the A.C. 14 has prefabricated flukes with increased surface area. Such a construction provides the increased holding power, weight for weight. The design of the early 'Bruce Anchor' combined the hook effect of the Admiralty Pattern and the spade effect of the stockless, to provide a high holding power anchor with no moving parts. This was widely engaged in the offshore industry. Its main disadvantage, because of its curved shape, was that it was difficult to stow when not in use.

Flipper delta anchors – Probably one of the most modern designs in anchor operations today. It is a high holding power anchor where the angle of the flukes can be changed and set to a respective desired angle to the shank. This variable fluke angle would be determined by the nature of the holding ground. It has become popular in the offshore environment. A tripping pennant is used to break the anchor free, prior to recovery.

Mooring anchors – Many and varied in designs. They are extensively employed in holding patterns to secure buoys and offshore floats. Usually a minimum of three coupled with a chain swivel unit is normal practice for holding a buoy or light float in position.

NB. There are numerous types of anchors current and past, employed within the marine industry and the readers interest is directed to the author's publication: Anchor Practice – A Guide for Industry.

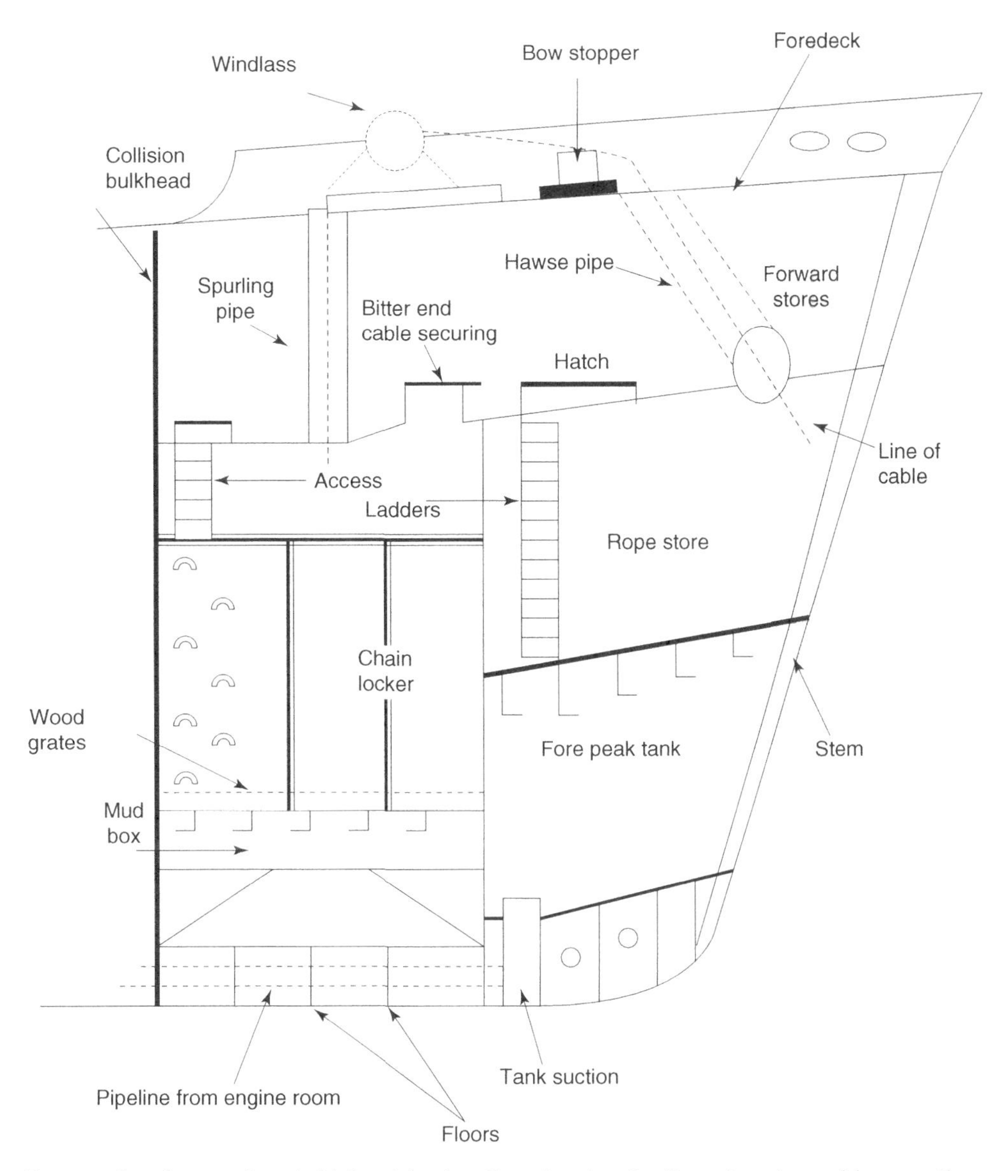

Bow section (conventional ship) – side elevation showing the line of anchor cable operation.

Windlass and anchor arrangement

The design of modern vessels, especially wide beam ships, has moved more towards separate windlass arrangements as opposed to a single centre line windlass. Separate windlass systems are suitable for Roll On–Roll Off ferry type ships which carry a 'Bow Visor' system, where a centre line windlass would be inappropriate.

Hawse pipes are set well aft to permit adequate clearance for any visor operation and the design generally means that the hawse pipes themselves are at a steeper lead, compared to the conventional anchor stowage designs.

Single anchor windlass, incorporates an end warping drum. Anchor securings are by means of the guillotine bow stopper (seen open), the windlass band brake on the gypsy and additional wire/chain lashings to pass through the cable.

Securing the 'Bitter End'

Current regulations require that chain cable can be slipped from a position external to the locker and the bitter end attachment is achieved by a tapered draw bolt system or other similar slip. This version (below) is seen where the cable passes through the side bulkhead of the chain locker.

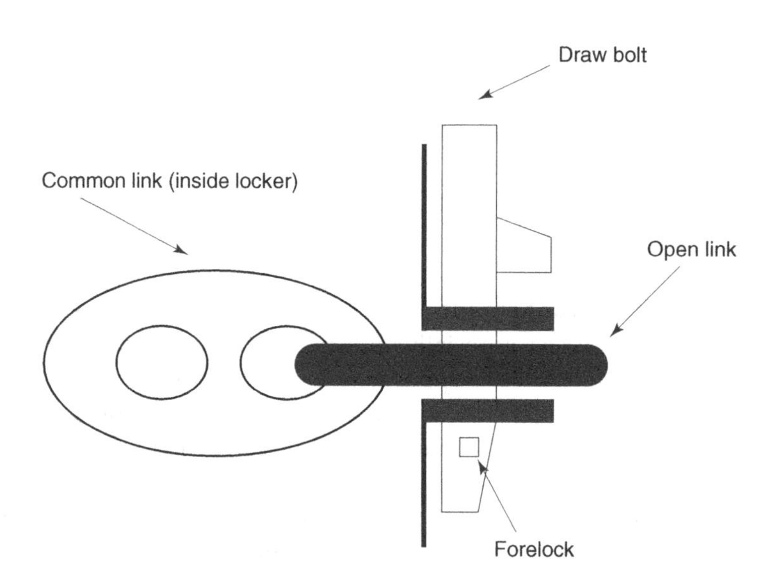

Releasing the bitter end of the anchor cable while the vessel is in active service would be considered an extreme action. Mainly because without two working bow anchors, the vessel would be considered un-seaworthy for the purpose of marine insurance. The cable would only be released by this method if the ship was in a dire emergency situation. However, it should be realized that the slipping of the bitter end is a regular feature of many dry dock operations and is usually carried out to allow the chain locker to be cleared for maintenance purposes and/or for inspection of the anchor cable itself, including the slipping arrangement.

The anchor plan

An anchor plan should be established between the interested parties, namely: The Ship's Master/Captain or Offshore Installation Manager (OIM), the Officer in Charge (OiC) of the anchor party, or the Master of Anchor Handling Vessel (AHV).

It would be expected that these key personnel would inform relevant crew members through an established chain of command, regarding relevant criteria. In the construction of any anchor plan the following items must be worthy of consideration:

1. The intended position of anchoring of the vessel.
2. The available swinging room at the intended position.
3. The depth of water at the position, at both High and Low water times.
4. That the defined position is clear of through traffic.
5. That a reasonable degree of shelter is provided at the intended position.
6. The holding ground for the anchor is good and will not lend to 'dragging'.
7. The position as charted is free of any underwater obstructions.
8. The greatest rate of current in the intended area of the anchorage.
9. The arrival draught of the vessel in comparison with the lowest depth to ensure adequate underkeel clearance.
10. The choice of anchor(s) to be used.
11. Whether to go to 'single anchor' or an alternative mooring.
12. The position of the anchor at point of release.
13. The amount of cable to pay out (scope based on several variables).
14. The ship's course of approach towards the anchorage position.
15. The ship's speed of approach towards the anchorage position.
16. Defined positions of stopping engines, and operating astern propulsion (Single Anchor Operation).
17. Position monitoring systems confirmed.
18. State of tide ebb/flood determined for the time of anchoring.
19. Weather forecast obtained prior to closing the anchorage.
20. Time to engage manual steering established.

When anchoring the vessel, it would be the usual practice to have communications by way of anchor signals prepared for day and/or night scenarios. Port and harbour authorities may also have to be kept informed if the anchorage is inside harbour limits or inside national waters.

NB. *Masters, or Officers in Charge, should consider that taking the vessel into an anchorage must be considered a Bridge Team operation.*

Bringing the vessel to a single anchor

It would be normal procedure for a ship's master to consider the approach towards an anchorage, and discuss the operation with the Officer in Charge of the anchor party, namely the Chief Officer of the vessel. Probably the most common of all uses of anchors is to bring a vessel into what is known as a 'single anchor' where the ship has adequate swinging room to turn about her one anchor position, with the turn of the tide and/or influence of the prevailing weather.

A planned approach to the intended position should be employed with the Master or Marine Pilot holding the 'con' of the vessel. The anchor party, on the orders of the Master, should clear away the anchor lashings and 'walk back' the intended anchor for use in ample time, before the vessel reaches the anchor position. The readiness of the anchor to be 'Let Go' should be communicated to the bridge by the intercom/ phone system or 'Walkie Talkie' radio.

The Master would turn the vessel into a position of stemming the tide and manoeuvre the ship towards that position (as per plan) where he intends to let go the anchor. By necessity, the ship will still be making 'headway' in order to attain this position. Headway is taken off at this point by using astern propulsion but it will be noticed that sternway will not take an immediate effect (masters will have to estimate when the vessel is moving astern and this is not always readily observed. One method is to sight the wake from the propeller moving past the midship's point towards the forward part of the vessel. This is a positive indication that sternway is on the vessel).

Fundamental principle of anchoring, is that it is the weight of cable and the lay of the 'scope' that anchors the vessel successfully, not just the weight or design of the anchor.

Once sternway is positively identified on the vessel, and the position of letting go the anchor is achieved, the Master would order the anchor to be released. The astern movement of engines would be reduced to an amount that the anchor cable could be payed out on the windlass brake, as the vessel continues to drop astern, slowly. The Officer in Charge of the anchor party would check the run of cable by using the gypsy braking system in order to achieve a lay of cable length along the sea bed. The Officer in Charge would endeavour not to pile the cable in a heap on top of, or close to, the anchor position. As the pre-determined amount of cable to be released is achieved, the engines should be stopped from moving astern. The cable will have been allowed to run and the brake would then be applied to check the amount of scope. This should serve the purpose of digging the anchor into the sea bed and stop the vessel moving any further astern, over the ground. The ship is described as being 'Brought Up' to her anchor and it would be the duty of the anchor party officer to determine when the vessel is 'brought up' and not dragging her anchor.

Amount of anchor cable to use (single anchor)

The experience of the Master will always influence the amount of anchor cable to be employed for a single anchor operation. Most masters would work on the premise that 4 × Depth of water would be considered as the absolute minimum. The nature of the holding ground, the range and strength of tide, the current and expected weather conditions will all be factors that influence the optimum scope.

The intended time period of staying at anchor would be a further factor. When all the variables are considered the Master would still probably add another shackle length for luck and ship security.

Clearly the available swinging room must reflect the scope of cable and keep the vessel clear of surface obstructions. Consideration of the amount of cable to use would be made well before the approach is made to the anchorage; the amount being established following a chart assessment of the intended anchorage and an assessment of all variable factors which could affect the safety of the vessel. The use of a comprehensive anchor plan in the form of a checklist could be seen as beneficial and is considered good practice within some shipping companies.

Swinging room – vessel lying to a single anchor

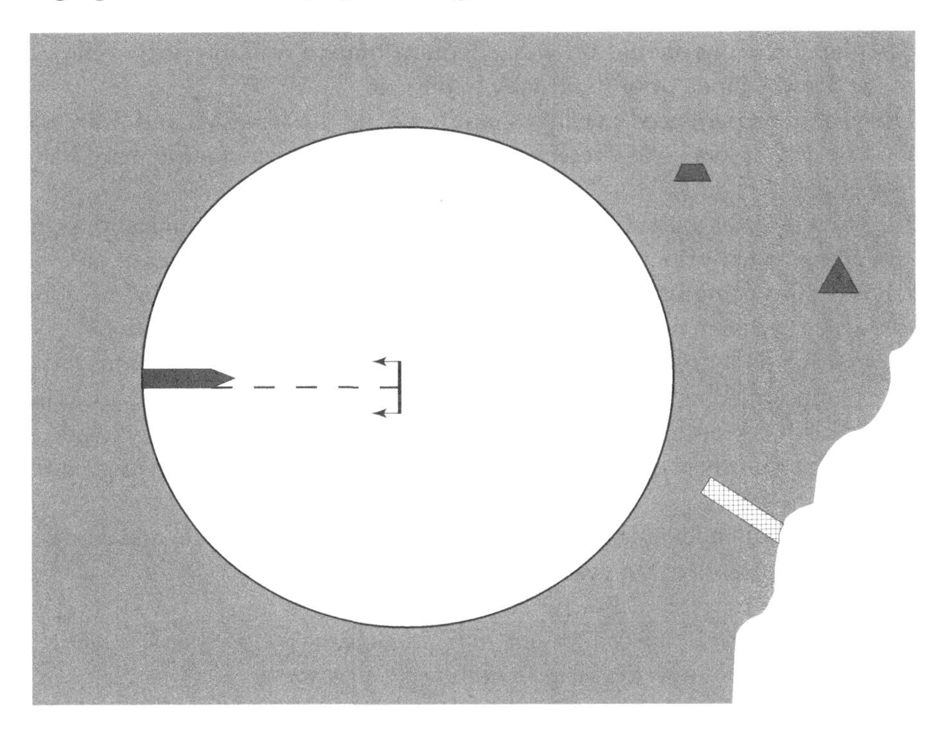

Surface obstructions must be significantly clear of the swinging circle, e.g. piers, buoys, navigation marks, etc.

Swinging room for a vessel at the single anchor will occur at the maximum scope of cable when at long stay. This circle of swing could be practically reduced by employing two anchors in the form of either a running or standing moor. Although these moorings are not generally common they are suitable when a large swinging circle is not permitted like within a canal or river, where sea room is restricted.

The vessel will swing through 180° with each turn of the tide (usually about every 6 hours). Movement of the vessel at anchor will also be influenced by the direction of wind. It is significant that wind over tide produces a powerful effect on the cable and, depending on the nature of the holding ground, may cause the anchor to break out and allow the vessel to drag her anchor, an extremely undesirable situation.

Watch Officers should be cautious to any traffic movement within the circle of swing, especially of traffic attempting to cross the ship's bow. Such traffic would be directly affected by the same direction of tide/current and be caused to set down on the anchor cable.

Operational safety when anchoring

Certain precautions when anchoring may seem obvious to the experienced seafarer. However, when dealing with five (5), ten (10) or twenty (20) tonne plus anchors, complacency can be the seaman's worst enemy. Routine operations should therefore include the following items:

1. Always check that the overside surface area is clear of small craft or other obstructions under the flare of the bow, at the intended area of letting go the anchor.
2. Routine operations should provide adequate time to walk the anchor back clear of the 'Hawse Pipe', prior to actually letting go.
3. Designated, experienced persons should operate the windlass and braking system. They should also be protected by suitable clothing including 'hard hat' and 'eye goggles'.
4. All parties to the operation should have inter-related communications. These should be tested prior to employing the ship's anchors. In the case of 'walkie-talkie' radios being used, these should operate on a clearly identified shipboard frequency and seen not to interfere with other local shipping operations.
5. The marine pilot or ship's Master who has the 'con' of the vessel should be continually informed as to the 'Lead of Cable' and the number of shackles in use. It would also be expected that the Officer in Charge of the anchor party would keep the bridge informed of any untoward occurrence, e.g. fouled anchor or windlass/power defects.
6. All recognition and sound signals should be employed promptly and correctly to highlight the status of the vessel.

NB. *It is expected that, when anchors are to be deployed, a risk assessment would be conducted prior to involving personnel and operating deck machinery.*

In the event that overside work is required in conjunction with the anchor operation, a 'permit to work', must be completed in addition to the risk assessment.

The watch at anchor

It should be clearly understood by any and all watchkeeping personnel, that when the vessel goes to anchor, she is still considered 'at sea'. As such, an effective and proper lookout must continue to be kept from the navigation bridge. The officer of the anchor watch will be responsible directly to the ship's Master for the well being of the ship, and should be familiar with the two greatest dangers, namely:

a) own ship dragging anchor or
b) another ship dragging towards own ship.

In either of these cases the Master would be expected to come to the bridge and take the 'con' of the vessel.

Watch duties, inclusive of keeping the lookout, would expect to include monitoring the performance of the weather particularly closely; keeping a listening watch for radio traffic and ensuring that the vessel displays the correct navigation signals in all states of visibility. Where small launch or tender traffic is in attendance, the monitoring of movements of such traffic is considered good ship keeping practice.

Other hazards also inherent to anchoring – and something the diligence of watch personnel can go some way to defend against – are: fire, piracy, collision from another vessel, pollution, dragging and shifting position.

NB. *Similar duties would be expected if the vessel is moored up to mooring buoys.*

Anchoring principles

The amount of anchor cable employed has always been considered the critical factor when bringing a vessel into an anchorage. The anchor itself acts as a holding point from which the cable can be laid in a line on the sea bed. Ideally, this line should be at a narrow angle from the sea bed to generate a near horizontal direction of pull on the anchor; the position lending to the term 'Long stay'.

Short stay is usually where the cable is at an acute angle to the surface and such a deployment would have a tendency to pull the anchor upwards, possibly causing it to break its holding of the surface at the sea bed.

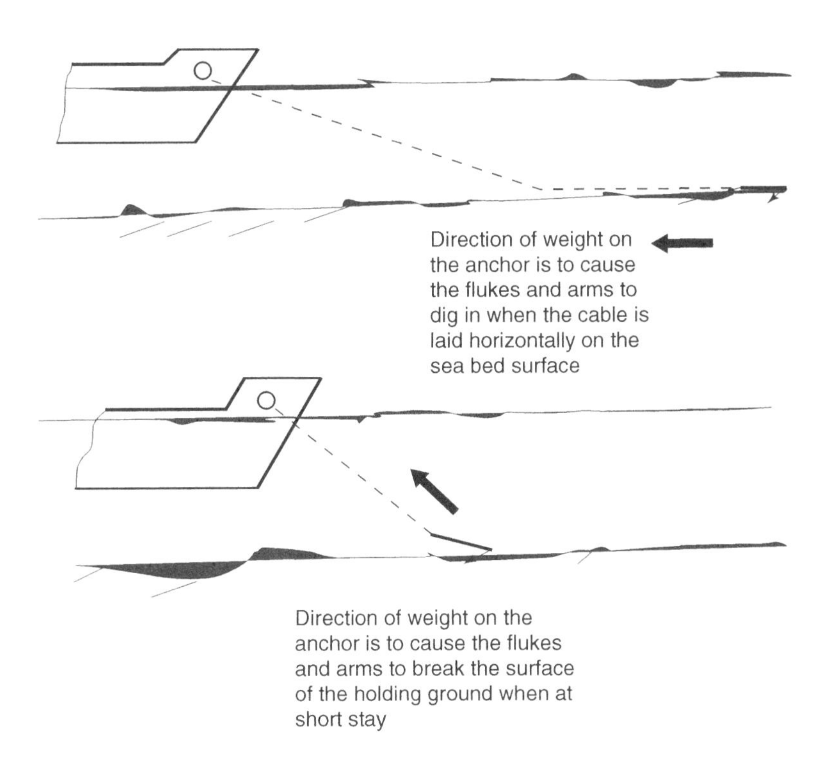

Detection of dragging anchor

One of the fundamental principles of the anchor watch is to ensure that the vessel does not break her anchor out and drag away from the anchor position. To this end, the weather conditions, state of currents and tides should be continuously monitored throughout the watch period.

Normal procedure for the watch officer at anchor would be to regularly verify the ship's position. Where dragging is suspected, the ship's position would be expected to change.

Such movement may be ascertained by any or all of the following methods:

1. Check the anchor bearings of the fixed landmarks. These references should be retained on the chart during the period of the anchorage; they should also be entered in the ship's deck logbook. If they are changing, the ship's position is changing and the vessel must be assumed to be dragging.
2. Obtain an immediate positional check from the GPS operation, to ensure that the instrument co-ordinates correspond to the Latitude and Longitude of the ship's anchored position. Any discrepancy in position, the vessel must be assumed to be dragging its anchor.
3. Engage the variable range marker of the ship's radar onto a fixed land object. If the range between ship and landmark opens or closes then the vessel can be assumed to be dragging its anchor.
4. Direct observation and hand contact with the anchor cable may give further indication that the ship is dragging its anchor. A dragging anchor would usually generate excessive vibration through the length of the cable, which could also indicate dragging (depending on the nature of the holding ground).

NB. Watch officers should not leave the navigation bridge unattended and, if checking anchor cable by this method, should wait to be relieved by the master or another watch officer.

5. A hand lead over the bridge wing with the lead on the sea bed. If the vessel was dragging its anchor the lead of the line to the lead would stretch forward towards the position of the anchor, indicating that the ship was dragging its anchor.
6. The use of beam transit bearings is also considered as a good indicator that the vessel may be dragging her anchor. However, the use of transits alone should not be accepted as being totally reliable, and would normally be used in conjunction with other methods of ascertaining movement in the ship's position.

Vessel dragging towards own ship

(It must be anticipated that the ship's Master would take the 'conn of the vessel', for such an incident.)

It is not unusual for vessels to share a common anchorage, and the additional concern for any Master or watch officer would also be to monitor movement of other ships in and around the anchorage. Where weather or tidal conditions are tenuous the possibility always exists of another ship breaking its anchor out and dragging towards your own position. Such an occurrence would be somewhat out of your control and the very least that an individual can do is raise the alarm and draw attention to the other vessel(s) of the developing situation. There is no one affective method to draw the attention to a vessel dragging its anchor. The scenario may be in daylight or during the hours of darkness and as such the methods to highlight the danger must include any or all of the following:

a) Raise the effected vessel 'by name' on the VHF radio, to advise of the movement (VHF should not be employed without station identification).
b) Give five or more short and rapid blasts on own ship's whistle.
c) Give five or more short flashes on own ship's signal lamp, directed to the effected ship.
d) Display the signal 'RB 1' in the international code to signify: 'You appear to be dragging your anchor' (by day).
e) Employ the ship's searchlight, directed towards the vessel dragging her anchor. Used in conjunction with Regulation 36 of the COLREGS.

Action by own vessel when a ship is seen to be dragging towards your own position would be dependent on the time available to take action. The ship's engines would be expected to be on 'stop' and ready for immediate manoeuvre, should they be required. Steaming over your own anchor cable may become a necessity to avoid contact with the vessel dragging. Alternatively, placing the rudder hard over could give the vessel a sheer to angle one's own ship, away from a position of contact.

Cable operations of either shortening or veering ones own ship's anchor cable may provide sufficient clearance to avoid contact. To this end an anchor party should be placed at a state of readiness as soon as the situation is realized.

Should the incident occur in conditions of poor visibility, vessels may also give an additional signal of 'one short, one prolonged, one short' blasts, to give warning of her own position to other vessels.

Your own vessel is faced with limiting options as the weight of anchors and cables deployed at the fore end will generally restrict vessel movement. An astern motion would also place excessive strain on cables and steaming over the cable would probably be seen as a preferable option, if considered necessary to engage engines to avoid contact.

Dragging the anchor

Once anchored and 'brought up', the main danger for the vessel's security is probably from dragging her own anchor(s). This detrimental situation is virtually always caused by a change in the natural weather conditions or current/tide changes, assuming that the vessel was correctly anchored in the first place. Dragging would normally be detected by the Officer of the Watch, at anchor and standing orders would invariably include instructions to call the ship's Master as soon as the movement is confirmed.

Many ship handlers have a policy to cater for a vessel dragging her anchor and the usual form would be for the scope to be increased, with the view that the additional weight in cable will cause the anchor to dig itself in again. Alternatively, a second anchor may be used, at short stay, to provide added weight at the fore end and reduce pitch on the vessel. The scope on the second anchor could also be increased if the need arose and the vessel was observed to continue to drag anchor.

Any of these options are worthy of merit, especially if the weather is known to be abating, but this fact cannot be guaranteed at the onset. The disturbing influence, causing the vessel to drag could well be increasing and the actions taken to initially protect the safety of the ship could well work against the well being of the vessel.

The disadvantage of the above actions is that they all restrict the manoeuvrability of the vessel in the event that the weather conditions become so bad that the vessel is left with the only option but to heave up anchors and cables and run for open water. Adding additional scope to a single anchor will not necessarily stop the ship from dragging, and would most certainly increase recovery time when weighing anchor.

Letting go a second anchor, underfoot with or without increased scope, would cause further delay and may incur constraints on the windlass, especially if the handling gear is old. Recovering one anchor in bad weather conditions may prove difficult, while attempting to recover two anchors, plus cable, may become just too demanding. Masters would invariably have to resort to using the ship's engines to ease the weight on the cables and may find themselves restricted to recovering one anchor at a time.

Author's Opinion: Ship handlers should give due consideration to the prevailing conditions and the historical weather patterns of the area in which they are anchored. Rather than encumber the vessel with more cable or even second anchors, make the decision to 'weigh anchor', and run to either a more sheltered anchorage, or seek open waters and ride out the bad weather. If the terrain and geography permit, the 'lee of the land' may be sought and the vessel could steam up and down until the weather subsides.

NB. *In all cases of bad weather, the ship's engines should be kept available for immediate use and the weather conditions should be continually monitored.*

The Master should have the 'con' of the vessel and the anchor party should be retained on 'stand-by' while conditions give reason for concern.

Baltic and Mediterranean moorings

Baltic moor – onshore wind, no tide effects

1. Approach the berth parallel to the quay, with the offshore anchor walked out. A stern wire mooring in bights should be passed forward, secured by light lashings and shackled to the ganger length of the anchor cable (this may require a boatswains chair operation inside the harbour limits).
2. When the ship's bow position is opposite the middle of the berth, let go the offshore anchor. Engines should be dead slow ahead and the rudder amidships. The onshore wind will cause the vessel to move towards position '3' parallel to the berth. Pay out the anchor cable and the stern mooring wire as the vessel closes the berth.
3. Stop engines and check the anchor cable and the stern mooring. Send away fore and aft mooring lines to the quay.
4. Tension the shoreside moorings and lay to a taut anchor cable and stern mooring wire, off the quayside.

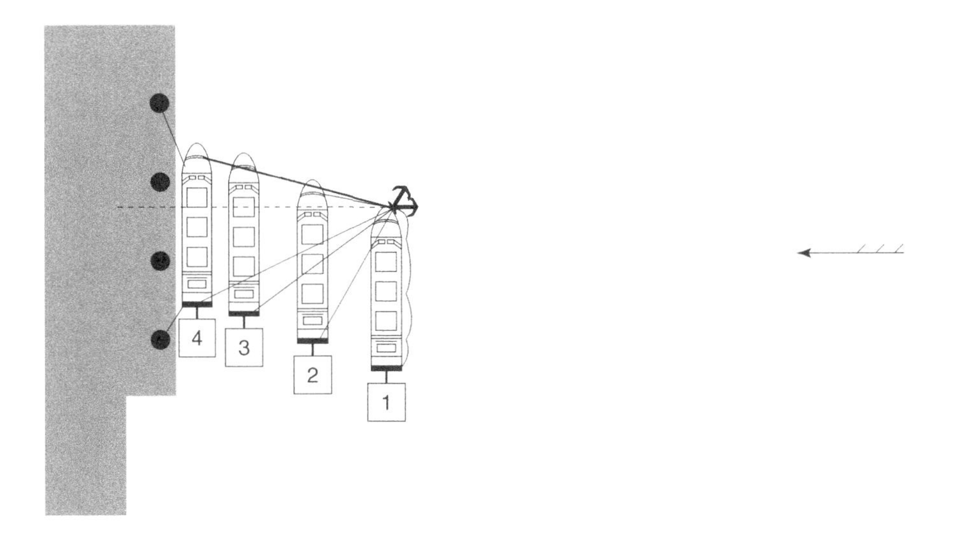

> ***Comment***: The purpose of the 'Baltic Moor' is to hold the vessel just off the quayside because the berth is concrete and probably unfendered. With an onshore wind, the vessel could expect to sustain shell damage if landing alongside such a quay. Alternatively, the quay could be a frail timber jetty and a heavy laden ship may cause damage to the quay itself. The shore side moorings are stretched to prevent the vessel ranging while laying parallel to the berth.

NB. The angle of approach to the berth may need to be more acute with vessels which have accommodation all aft.

Mediterranean moor

The objective is to moor the vessel stern to the quay by means of mooring ropes aft and the use of both of the ship's bow anchors forward. This type of mooring allows more vessels to moor to the berth where there is limited quay space. Some specialist vessels like tankers and Ro–Ro vessels also use the arrangement for stern discharge via an aft manifold or stern door, respectively.

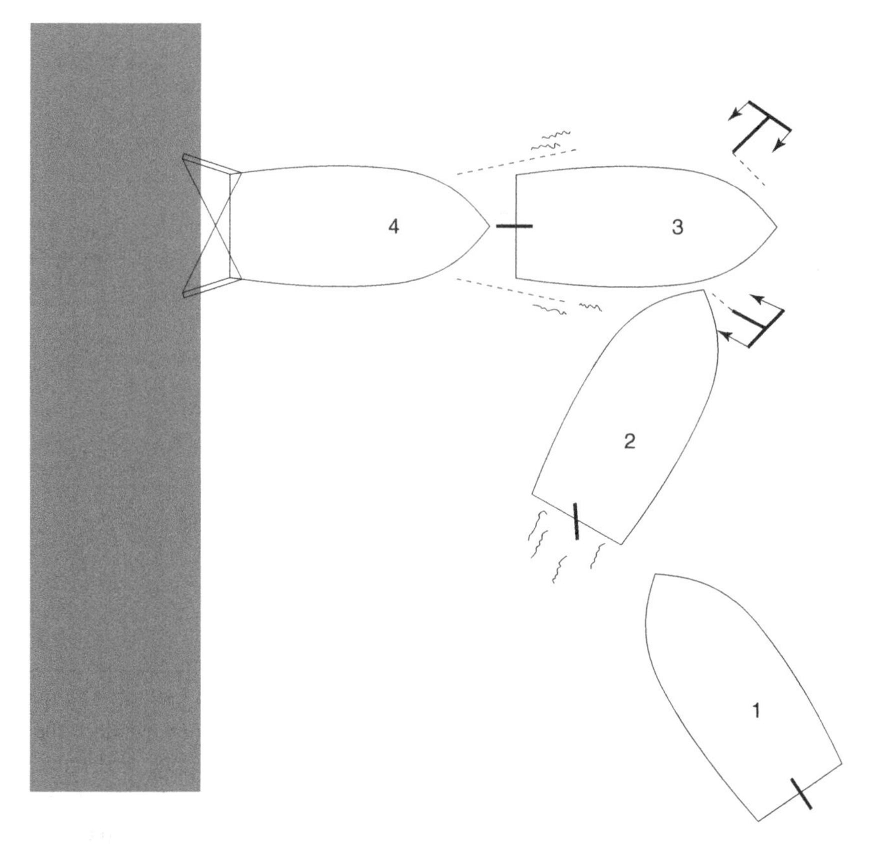

Mediterranean Moor – right hand fixed propeller.

Mediterranean moor procedure

1. Approach the berth port side to, at about 1½ ship's lengths distance off. Dead slow ahead. Cant the bow towards the berth.
2. Hard to starboard and let go the starboard anchor. The starboard helm will cause the stern to swing towards the berth.
3. Stop engines. Full astern and let go the Port Anchor, rudder amidships. The effect of transverse thrust will keep the stern swinging towards the berth.
4. As the vessel gathers sternway, stop engines and pay out on both anchor cables. When the vessel reaches within heaving line distance off the berth, check the cables and pass quarter lines and crossed inboard springs.

NB*. Once the moorings are all fast aft, these can be tensioned by giving a slight heave on both anchor cables. Scope of cable on both anchors is limited to provide a short stay on both anchors, once the vessel is stern to the quay.*

Comment: In order to complete the manoeuvre with no tide as such in the Mediterranean Sea, it is essential that the vessel with a right hand fixed propeller makes a Port side to approach, in order to maximize the effects of transverse thrust.

The uses of the Mediterranean moor

The main reasons for the Mediterranean moor are to optimize the available space on the berth. However, once achieved, it provides some distinct advantages for stern discharge methods but some disadvantages are present, in that the use of shoreside cranes is denied and the cargo vessel would have to use its own ship's lifting gear.

It should also be noted that the vessel is more exposed as when compared to a vessel which is fully secured alongside. With the direction and aspect of the mooring, personnel going ashore would ideally require a boat (not in the case of a Ro–Ro with stern ramp/quay contact).

The 'Superfast II', passenger/vehicle ferry manoeuvres to establish into a Mediterranean Moor alongside the passenger/vehicle ferry 'Express Penelope'. Both vessels will be secured by stern moorings with both anchors released at short stay.

The Mediterranean moor

When a vessel is equipped with enhanced manoeuvring aids like Controllable Pitch Propellers together with 'Bow Thrust', the approach to the mooring can be made with either a Port side or Starboard side to quay approach, at a distance of about one and a half ship's lengths.

Both anchors should be walked back clear of the hawse pipes and held on the windlass brakes. Once the centre position of the intended berth is reached, the way should be taken off the ship, prior to commencing a turn in the offshore direction.

The inshore engine should be placed ahead, with the offshore engine placed astern. Maximum bow thrust 100 per cent should be given in order to turn the vessel about the midships point (Position 1).

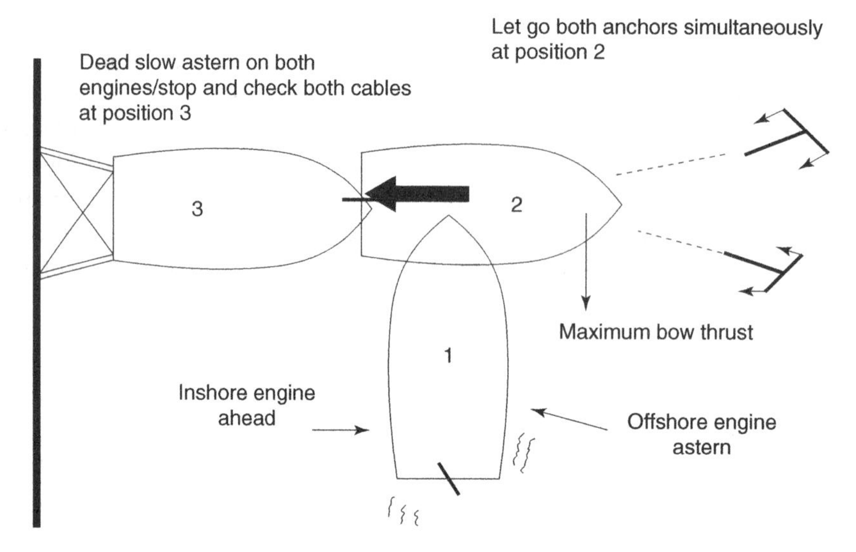

At position '3' the vessel should be within heaving line range of the quayside and moorings can be passed to secure the stern, once all fast aft, tension cables at the fore end to render stern moorings taut.

The use of an anchor buoy

An anchor buoy is rarely used these days, other than in the Offshore Industry. The principle of its use was to identify the position of the anchor and/or aid recovery in the event that the anchor was lost. Many ships used to carry a plastic buoy specifically for this purpose, but a wood float or sealed oil drum often performed the same function.

The buoy line is meant to be attached to a convenient position on the anchor, ideally to the anchor shackle or the links of the ganger length (unless a gravity band is fitted). The securing is of concern, as there is a risk of parting the wire with the action of the arms and flukes as they angle to dig in.

The choice of material for use as a buoy line or pendant would depend on the intended function and the length of stay. Some buoy lines may be a combination of wire and polypropylene, especially if the vessel does not intend to remain at anchor for a lengthy period. Other buoy lines/pendants which may be in the water for a considerable time and/or used for recovery purpose (as in the Offshore environment) would inevitably be of all wire construction.

Preparation of securing the anchor buoy to the anchor is usually carried out as the vessel approaches the anchorage, the buoy line being passed overside and secured onto the anchor. The anchor may or may not have already been walked back, clear of the hawse pipe. Adequate slack on the buoy line would be held in bights by sail twine, so that when the anchor is let go the line is carried away with the anchor.

The reader will appreciate that the rigging of the buoy line overside is cumbersome, and its recovery when heaving the anchor home could also cause inopportune problems. This is probably the main reason why the practice has been virtually discontinued in the shipping industry, although buoying salvage sites is still widely employed.

In the event that an anchor or cable is intended to be abandoned, then the anchor buoy and line would need to be substantial to aid recovery at a later date. Any ship's

Master, given the time and the circumstances, would anticipate a recovery operation and the use of a 24 mm flexible wire rope, held by a substantial buoy, would be appreciated by salvage operators (for average weight of anchors).

Running and standing moors

The running moor

The objective of the manoeuvre is to moor the vessel between two anchors with restricted swinging room. The manoeuvre is employed in tidal rivers, canals or harbours where sea room is limited and swinging room must be reduced. Once completed, watchkeepers must monitor any change in the wind direction which is likely to cause a foul hawse condition.

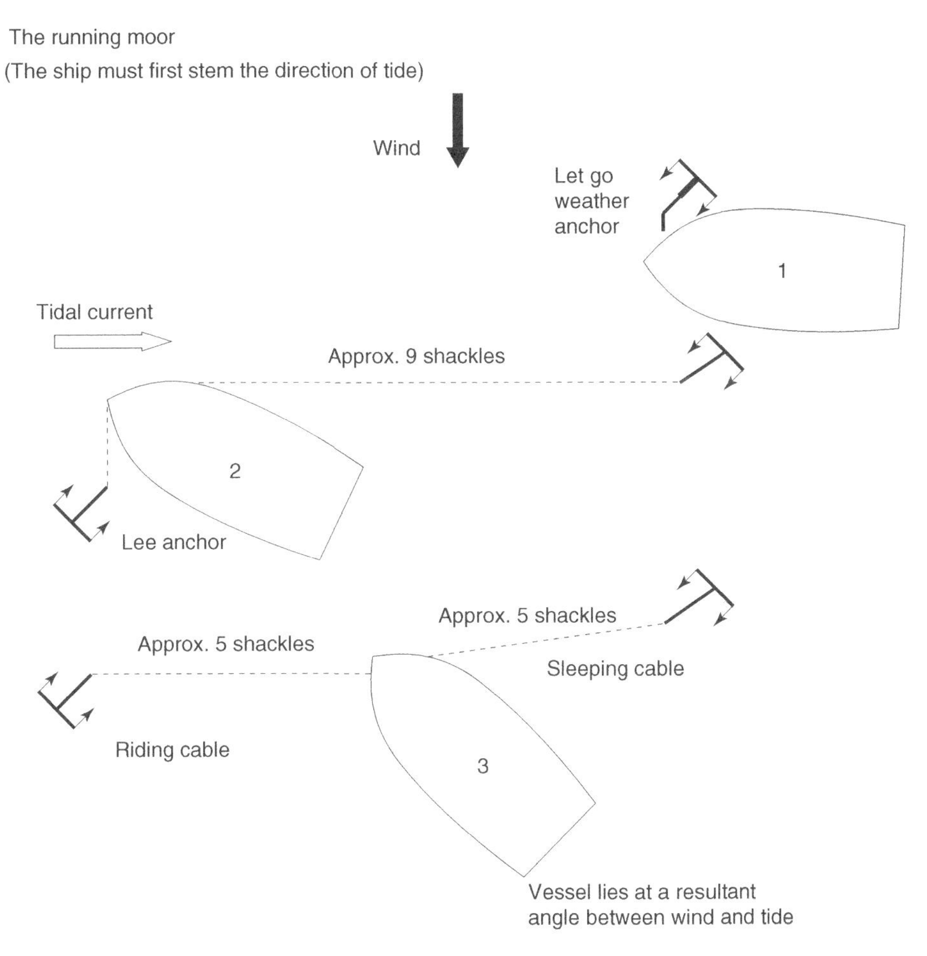

The running moor (the ship must first stem the direction of tide).

Running moor – procedure

1. The vessel should stem the tidal flow with both anchors cleared and on the brakes ready for deployment. It is essential that the windward anchor is let go

first in the running moor operation. The vessel is then allowed to run on, usually at dead slow, on engines into the tidal flow towards position '2'.

2. Let go the windward anchor (Starboard anchor) and continue to let the vessel run on, paying out the cable on the anchor for approximately nine shackles.
3. Stop the engines having paid out about eight shackles. By the time nine shackles have paid out, the vessel should be almost stopped moving ahead. As the tide effect takes the vessel astern let go the leeward anchor (Port anchor). Engage the windlass gear on the starboard anchor and shorten cable as the vessel drops astern, while paying out on the Port anchor.
4. Once the cables are at five shackles to each anchor, stop engines and stop windlass operations. The vessel will then be seen to lie to five shackles on the Port (Riding Cable) and five shackles on the starboard (Sleeping Cable).

NB*. The vessel would be expected to turn through 180° with each turn of the tide. The danger with two anchors deployed occurs if a wind change takes place, causing the cables to cross and end in the foul hawse condition.*

The reader might feel that the scope of five shackles on each anchor is not compatible with the movement of paying out an overall nine shackles. The riding cable will in fact be stretched at long stay, taking the weight, while the sleeping cable will have four shackles on the sea bottom with one shackle rising to the hawse pipe.

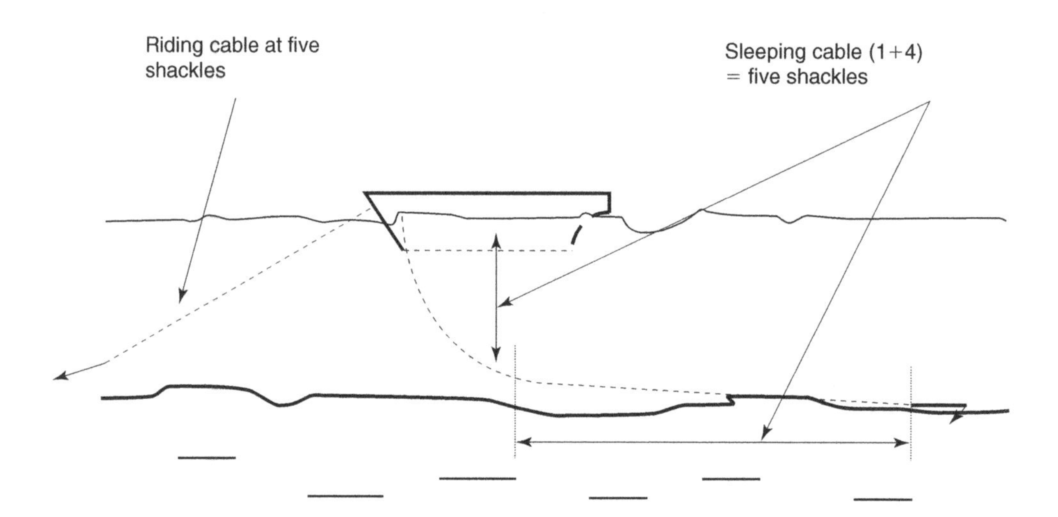

The standing moor

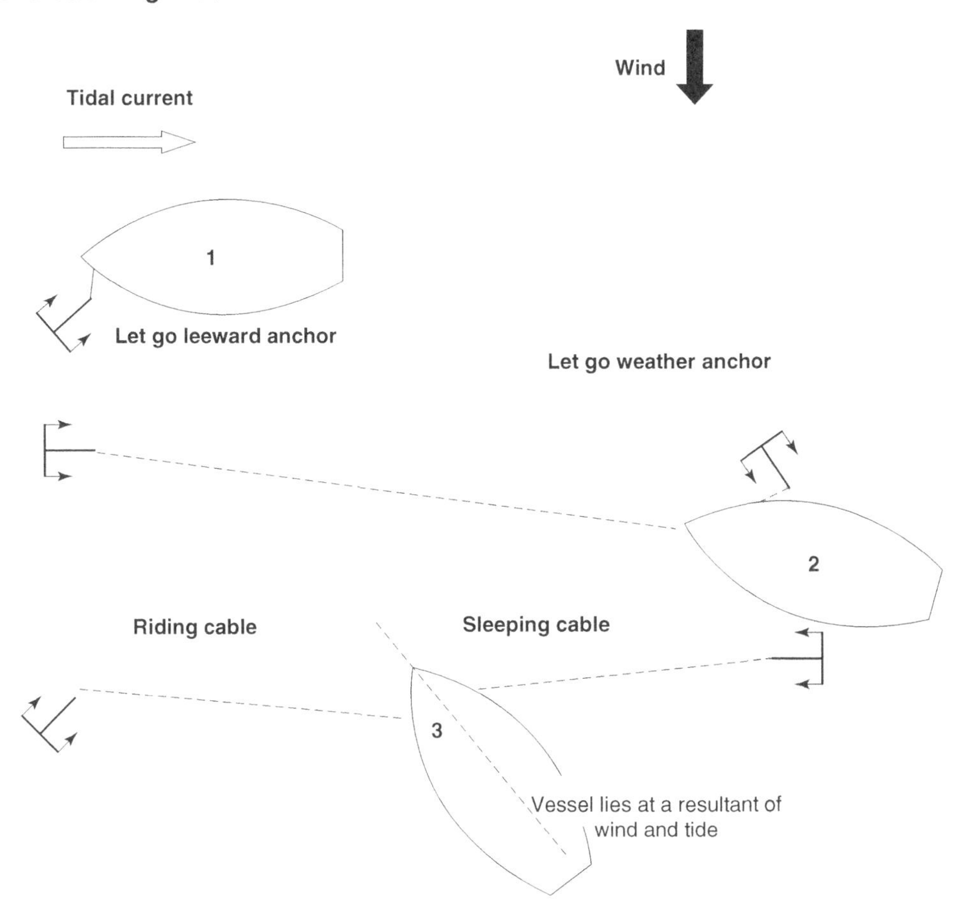

The standing moor (vessel must first stem the direction of tide).

The standing moor – procedure

This manoeuvre establishes the same mooring scenario as with a 'running moor' in that the vessel is moored between two anchors with reduced swinging room. The method of achieving a standing moor is similar, but is noticeably different by its procedure.

1. Stem the tide as in position '1' with both anchors walked out. Pass over the intended mooring position by about five shackles' length of cable. Let go the LEE ANCHOR and pay out the cable as the tidal direction allows the vessel to drop astern to position '2', a distance of about nine shackles, down from the position of the deployed anchor.
2. With nine shackles deployed to the lee anchor, apply the windlass brake. Let go the weather anchor and engage the gear on the lee anchor already deployed. Shorten cable on this 'riding cable' as the vessel moves ahead while at the same time pay out on the weather anchor (now the sleeping cable) to bring the vessel to a position midway between both anchors.
3. The vessel should adjust cables to show equal length (five shackles) on each cable. The riding cable will then lie with five shackles at long stay into the tidal direction,

while the sleeping cable will lie with five shackles, without any weight bearing on the cable.

NB. The vessel will adopt a resultant angle of position taking account of the tidal direction and the direction and force of the wind.

Note this manoeuvre could in theory be carried out without using the ship's main engines as with, say, a disabled vessel, the windlass and tidal stream being allowed to provide the motive force to the vessel to move over the ground (this option is not possible with a running moor, which requires main engine propulsion).

> **Comment**: If either the running or standing moors are required, some masters would favour adjusting the scope of cable to respective anchors to ensure five shackles to the ebb tide and four shackles to the flood, as a minimum requirement.

Mooring to two anchors. The Class 1, Passenger Vessel, 'Seaward' lying to port and starboard bow anchors, off the coast of Mexico. This particular vessel is also seen with a centre line anchor in the stowed position on the stem. Centre line anchors are not usually carried in common practice, except for specialist tonnage as in the Offshore Industry.

The open moor

This type of mooring is carried out in non-tidal conditions, such as when in a fresh water river. The use of two anchors, one off each bow with approximately equal lengths of cable is carried out to give greater holding power against a strong directional flow.

The open mooring should not be confused with the alternative use of two anchors where a second anchor has been used for additional holding in bad weather (usually one being deployed at long stay, with a second anchor deployed later at short stay as the weather deteriorates).

The open moor is not a practical option for tidal waters as, once the tide turns, clearly the anchor cables would foul.

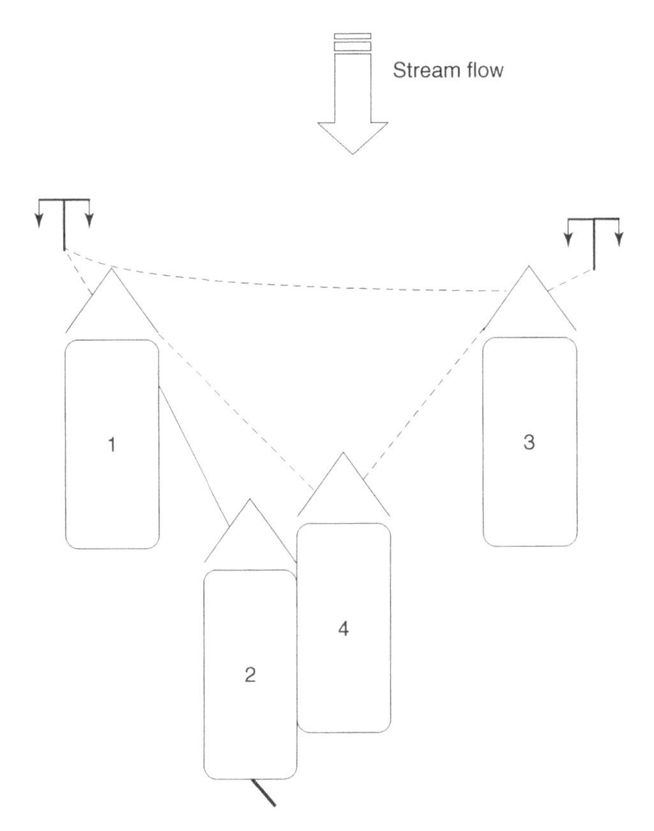

Use of main engines and helm to manoeuvre from position '2' towards position '3' will cause the cable to bellow out in a beam direction, prior to letting the second (starboard) anchor go at position '3'.

Each cable can be checked and shortened as the vessel falls astern between the two anchors, so that each anchor has approximately the same scope.

> **Comment:** It is emphasized that this type of mooring is for non-tidal conditions.

Open moor – procedure

1. Stem the current flow and adjust engine revolutions so that the vessel is stopped over the ground. Reduce the rpm and cant the bow to starboard. As the vessel moves astern and sideways let go the port anchor and pay out the cable to about four to five shackles.
2. Hold on to the port anchor and increase the rpm. The vessel will move ahead and sideways.

3. Reduce the rpm and bring the vessel head to current. Let go the starboard anchor and pay out the same amount of cable as on the port anchor.
4. Stop engines and bring the vessel up to four or five shackles on each anchor.

Dredging an anchor

On occasions it can be a practical option to dredge an anchor on the sea bed, usually carried out when approaching a berth. The objective here is to let the anchor go at short stay; definitely no more than twice the depth of water. The operation can take one of two forms, namely:

a) to use the cable to act as a spring effect, turning the aft part of the vessel in towards the berth (the pivot Point acting well forward), or
b) to position the vessel up tide, ahead of the berth, and allow the tide effects to move the vessel astern, deliberately dragging the anchor backwards, allowing the ship to fall back towards the berth (the rudder action being used to cause the vessel to draw to a position parallel, alongside).

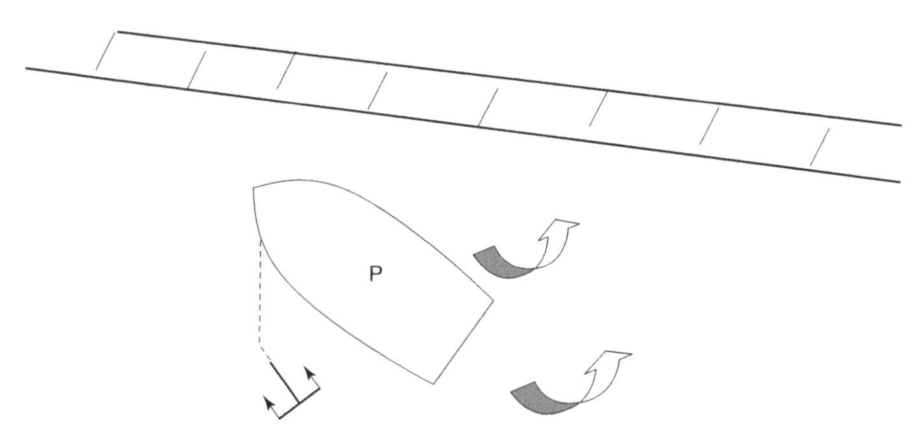

Dredging a single anchor to a spring effect on the cable. The offshore anchor is let go at short stay, as the vessel approaches the berth. As the run of the cable is checked, the spring effect, combined with rudder action, causes the stern to move towards the berth. The anchor would probably be left deployed in the up and down position, to assist when letting go and departing the berth.

Dredging an anchor from an up tide position is an alternative option when berthing. With a flow stream past the rudder, the helm can be applied to generate a parallel closing movement towards the berth.

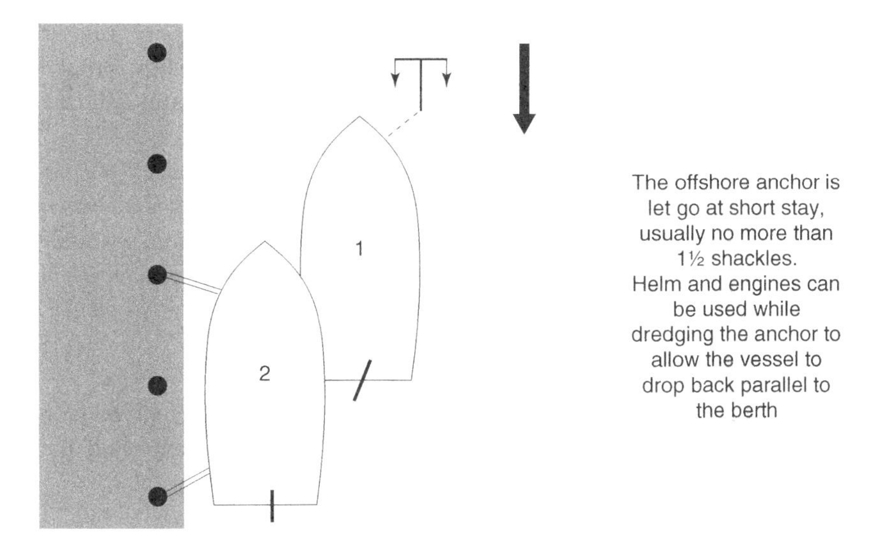

Dredging an anchor down to the berth, on a single right hand fixed propeller vessel. Once alongside and moored the anchor would normally be drawn home, being at short stay. If the anchor is left deployed, the cable would be walked back to the up and down position.

Anchoring large vessels, e.g. (VLCC)

When a large vessel like a VLCC intends to deploy anchor(s), the 'anchor plan' should include due consideration to the speeds of approach and the speed of the vessel over the ground when walking back the anchor over the intended anchorage position. A recommendation of 0.25 knots, over the ground, should be considered as appropriate but such an estimate could be influenced by prevailing weather conditions. It is also worth noting that monitoring the ship's speed at such a low value does have problems and may prove difficult even with updated equipment.

The ability of the vessel to retain such a recommended speed would lend to achieving a suitable lay of cable over the sea bottom, without placing undue accelerations on the mooring equipment. The capabilities of 'Band Brake Systems' on larger tonnage would already seem to be operating at their upper limits and any increase in the momentum of the running cable, caused by increased vessel speed, must be considered as undesirable. Any such increase could cause overheating of a braking system and result in lost anchors and cables.

In the event that unrestricted descent of the anchor is allowed to take place, i.e. letting go, then damage to the windlass gearing and/or motors may be unavoidable. Speed limiting devices operating on the band brake may make the operator's task of control easier but overheating could still become a problem with subsequent loss of braking efficiency. As such, over reliance on a restrictive speed, system should not become the order of the day, and the principle of walking the anchor back all the way should be adhered to.

It should be realized that +20 ton anchors are not unusual on large tankers and most masters would generally not wish to use anchors if the vessel could be safely allowed to drift in an area of clear water. Where an anchor is to be employed, turning the vessel into the tide would tend to take the 'way' off the vessel providing an opportune time to commence walking back on the cable. Once the anchor contacts the sea bed, the tidal direction would cause the vessel to drop away and astern from the anchor position.

***NB**: Large heavy anchors limit the types of anchor operations that can be carried out at sea by the ship's crew. The ships are usually limited with the type of lifting gear and the Safe Working Load of specific handling equipment, like shackles and wires. Invariably, where maintenance or specific anchor operations may be necessary, the service of a Dry-Dock would normally be required for such large vessels.*

An estimate of ship sizes was issued in the Seaways Magazine in 1983 (see below). This may provide some guidance as to the amount of scope expected for normal anchor use.

Ship size	*Scope, general estimate*	
Deadweight	*Loaded*	*Ballast*
20,000–50,000	7 × depth	9 × depth
50,000–90,000	7 × depth	9 × depth
Over 90,000	6 × depth	8 × depth

The amount of cable employed will always be influenced by the prevailing conditions.

A suggested approach to anchor a large vessel like a VLCC is shown opposite.

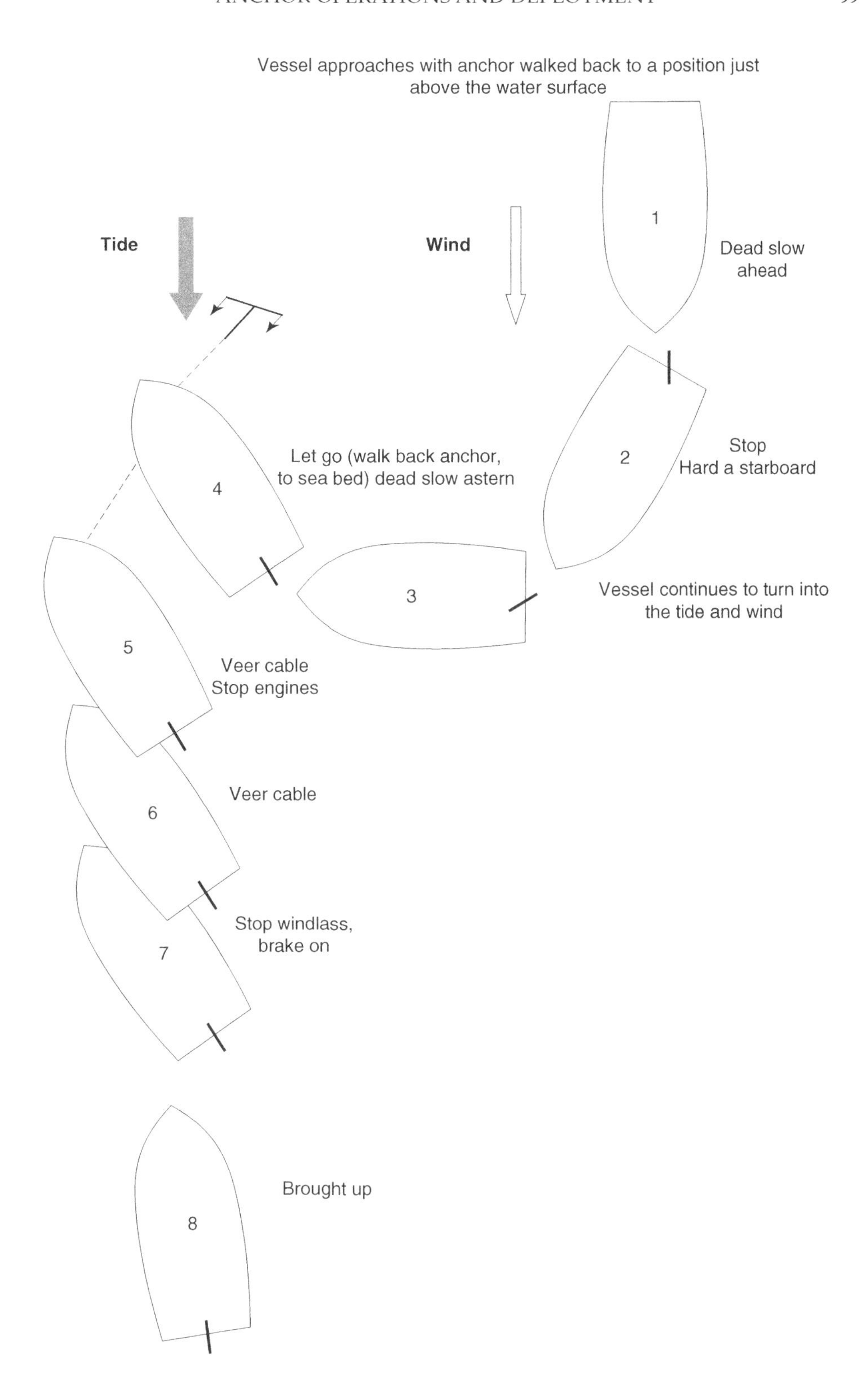
Vessel approaches with anchor walked back to a position just above the water surface
Tide
Wind
1
Dead slow ahead
2
Stop
Hard a starboard
3
Vessel continues to turn into the tide and wind
4
Let go (walk back anchor, to sea bed) dead slow astern
5
Veer cable
Stop engines
6
Veer cable
7
Stop windlass, brake on
8
Brought up

Emergency use of anchors

Vessels make use of their anchors as a matter of routine as and when required. However, there are occasions when they are required in emergency situations. Some examples of these could be experienced if the vessel had to beach and run into shallows, or hold off a 'lee shore' in the event of engine malfunction (not that anchors are designed to hold the total weight of the vessel).

Where a ship might have to break her own cables – as with a foul hawse – then such action would be considered as an incident which leaves the vessel vulnerable until the situation is rectified. Similarly, a vessel with a fouled anchor may not be able to deploy both bow anchors and, therefore, in certain circumstances may be considered unseaworthy, until the foul is cleared.

Anchor work operations are, by their very nature, heavy duty type activities for crew members and it is important that persons in charge of such incidents and emergencies are conscious as to the safety requirements that are necessary to go along with such demanding work. Risk assessments prior to carrying out any operation should be considered as standard practice.

Deep water anchoring

Taking a ship to anchor in deep water is not usually conducted as a matter of choice and Masters, Pilots and Officers in Charge of vessels intending to anchor should be concerned with anchoring in a safe manner. It is normal practice when anchoring, to employ a minimum scope of four (4) times the depth of water. Where excessive depth is present such a minimum may, by necessity, become increased, bearing in mind that a new windlass would have the capacity to lift 3.5 shackles of cable, plus the weight of the anchor, when in the vertical. Older equipment could expect to lead to a degree of lesser efficiency on the cable lifter.

In every case of deep water anchoring, the anchor MUST be placed in gear and walked back all the way to the sea bed assuming that it is at a lesser depth than 3.5 shackles. Clearly, any length over and above this could well fall outside the control of the braking system of the windlass and the capability to recover the cable length and the anchor. Under no circumstances should the anchor be 'Let Go'. Such action in deep water could well cause the brake system to burn out and leave the windlass without control.

NB. An exception to the above would only arise in the event that the vessel was in immediate danger of being lost. In which event, sacrifice of the anchor and cable length would be seen as an acceptable exchange for the well being of the ship.

Pointing the ship

When lying at anchor your vessel may be required to provide a lee for barges to work general cargo. This can be achieved by pointing the ship to create a lee for barges/launches to come alongside safely. This operation is achieved by running a stern mooring wire from 'bitts' aft, back up through the hawse pipe and shackling it onto the anchor cable.

The cable is then veered to provide the vessel with a directional heading off the weather and provides a lee for the operational use of barges or boats alongside.

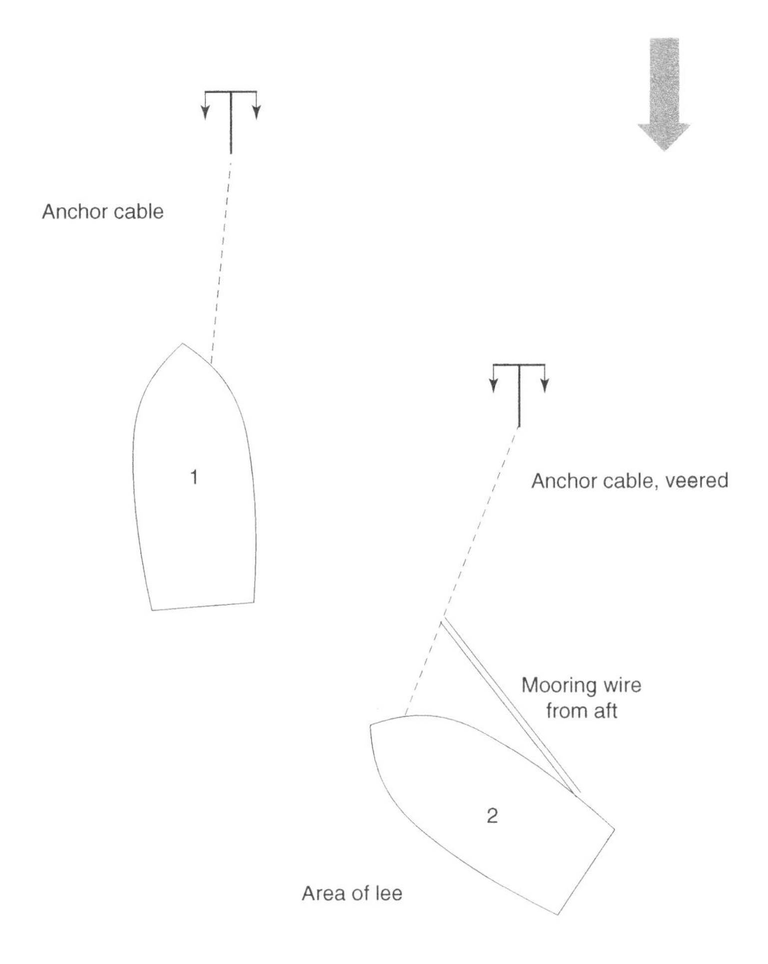

Anchoring the vessel in ice conditions

The general advice on deploying anchors when inside ice conditions is: do not anchor. Choose any other alternative, but do not anchor if another option is present. Clearly, no Ship's Master would willingly wish to put an anchor down in heavy ice, but it is not considered good practice in any ice condition.

The adversity to deploying anchors in ice stems from the experience of those who have tried in the past. Accumulation of surface ice between the hull and the anchor cable, where it breaks the surface, can be considerable. Large floes can become wedged and cause not only ice build up around the anchor cable but also down the side of the vessel.

If the vessel is nipped at one end, namely the bow, ice concentrations will not freely work themselves past the stern of the vessel and could accumulate down the ship's side. Such accumulations of possibly large ice floes along the side and in the vicinity of the cable could expect to generate concentrated weight against the ship. This weight in itself could cause the cable to part with excessive strain, or at the very least, come to play on the anchor and cause it to break out. Dragging anchor at any time is not a good experience, dragging anchor in ice conditions must be considered particularly hazardous.

Where anchors and windlasses are being employed in extreme cold climates it is always advisable to leave the windlass running during periods of non-use, to ensure machinery does not freeze up. Prior to entering ice regions, it is always considered prudent to ensure that pipelines are adequately lagged with insulation, especially pipes leading to deck machinery like mooring winches and windlasses.

Anchor parties, if required, are usually going to be exposed to extreme climates if the anchors are to be deployed. Such conditions, even when men are adequately clothed in protective, insulated wear, are not ideal and working conditions on slippery decks must be seen as a continuous hazard. If anchoring can be avoided, avoid it. You do not expose the crew, and the risk of problems with a fouled or frozen anchor is eliminated.

Alternative to use of anchors could be to revise route planning and/or adjust speed so as not to arrive at the necessity to anchor. A second alternative would be to steam up and down, hopefully in clear open waters away from ice infestation. A third option would be to consider a lay-by berth, but such an action could incur Port and harbour dues.

Use of mooring boats

Mooring boats are employed for many routine, as well as for the more unusual, operations. When engaged in running long drifts in mooring ropes, possibly from securing to non-conventional quaysides, it may be considered a routine duty for a mooring boat, compared to more unusual operations like mooring to single buoy moorings (SBMs) or passing slip wires to mooring buoys.

Whatever the task in hand, clearly operations with small boats, especially in poor weather conditions, must be considered somewhat hazardous for the personnel so engaged. Apart from the potential of interaction occurring between the parent vessel and the small craft, the exposed boat and the respective personnel are, in the majority of cases, very close to the water at a low freeboard. The working of the craft also tends to involve the handling of heavy warps and/or specialist equipment, in a confined deck area.

Personnel are also often expected to jump buoys for the securing of moorings, passing slip wires or securing anchor cables. Whatever the duty, personnel in this capacity, especially when expecting to 'jump the buoy', should always wear lifejackets. These would normally be available from the parent vessel if requested or if own crew members are involved. Positive communications should also be available at all times between the Bridge of the parent vessel, the mooring deck controlling officer and the mooring boat coxswain.

A mooring boat engaged in securing chain mooring recovery buoys for a shuttle tanker mooring to a Floating Storage Unit (FSU).

Non-standard mooring operations

Not all mooring operations are regular or conducted with standard mooring decks. Specialized vessels, certainly in the Offshore environment, are often engaged in mooring to SBMs or Floating Storage Units. Specialized moorings are usually employed to secure the ship prior to commencing cargo operations with pumping stations.

Moorings tend to be of a heavy variety combining heavy duty warp and/or mooring chains. A ship's mooring deck is likely to be fitted with Auto Kick Down (AKD) stoppers and special leads to accept chain moorings, while the operation of recovering the moorings is often by pick-up buoys. Special fitments on shuttle tankers are designed for securing the vessel safely for a limited time period to complete cargo and permit speedy release.

The offshore moorings from a large shuttle tanker to a Floating Storage Unit. The moorings are recovered via pick-up buoys with the chafe chains held by AKD stoppers, prior to passing the floating oil pipe seen at surface level.

Some of these ships often incorporate a forward conning position, remote from the bridge, to permit precise, close manoeuvring to the installation.

Mooring chain stoppers. A type of chain mooring arrangement for use when securing shuttle tankers or supply vessels, often used, known as an AKD stopper to hold the chain moorings. It is normal procedure to have a stopper to each chain, to port and starboard; mooring ropes may be an additional optional mooring.

Mooring to buoys – by anchor cable

Several ports around the world specifically require ships to moor up to buoys using the anchor cable of the vessel. It has to be said that this is not a common occurrence, but does happen from time to time. The ports in question are usually those that are regularly threatened by tidal surge or bad weather conditions often generated by monsoon conditions or within the Tropical Storm latitudes.

Many oil exporting countries also make use of the buoy mooring facilities offshore by bringing large tankers into SBMs or making connections to Floating Production Storage and Offshore systems (FPSOs). Specialized mooring arrangements secure the vessel, while floating pipelines are drawn on board to effect cargo transfer.

A chain mooring is sent away through a ship's lead to an SBM; once connected, the floating pipeline would be engaged to effect oil transfer.

Mooring to buoys (mooring lines)

The possibility that a ship may be required to moor between two mooring buoys is common practice in many ports of the world. The mooring may employ the use of the anchor cable or not, but more often employs the use of mooring ropes secured directly to the buoy. The basic needs require the use of a mooring boat, with a 'buoy jumper' wearing a lifejacket and additional securing equipment to join the rope to the ring of the buoy.

Methods of securing the ropes to the buoy vary but common practice is to send either a mooring shackle with the rope eye, or a length of about three (3) metres of rope lashing to secure the two sides of the eye, under the bight of the rope.

The vessel should be manoeuvred to bring the inner mooring buoy abeam of the break of the foc'stle head and off the port bow, passing a head line through a centre lead to the mooring boat for running and securing to the buoy. If a centre lead (bull-ring) is not fitted, two separate rope eyes, one from each bow, must be sent.

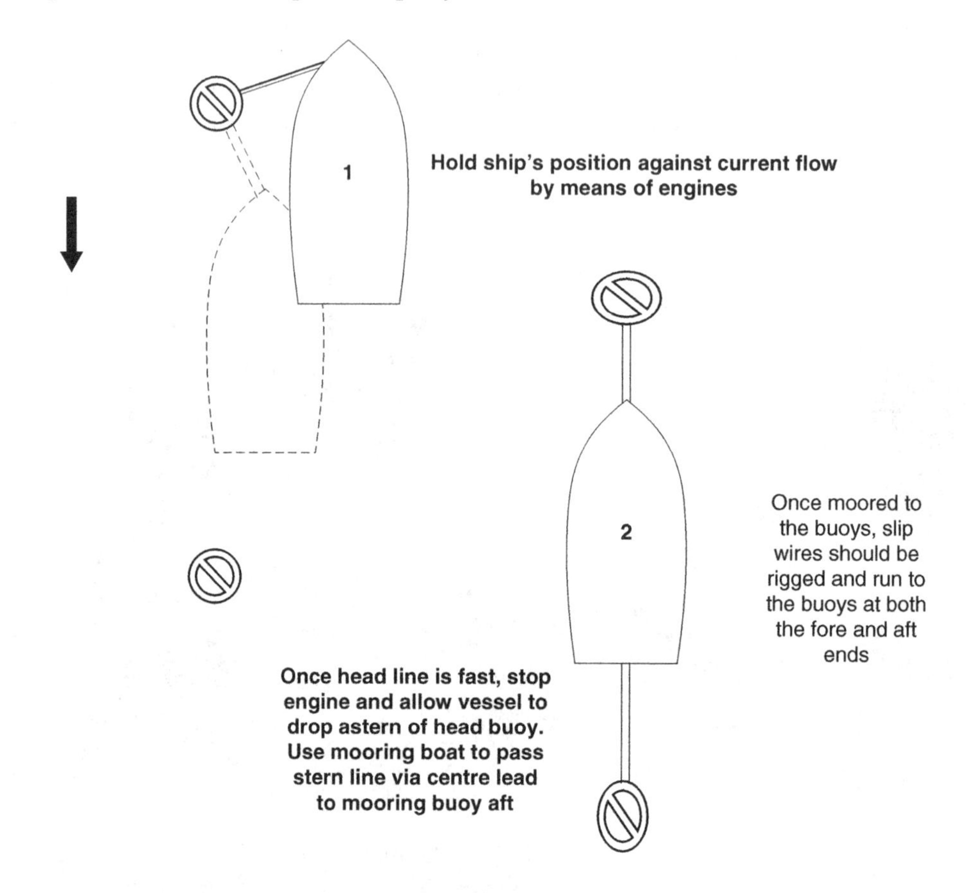

Mooring to buoys – procedure (tide ahead)

1. Stem the tide. If the tide is astern, then go past the buoys and turn the vessel short round to stem the tide. Position the vessel off the forward buoy. Use the engines to hold station. Pass all the forward headlines to the buoy.
2. Once the headlines are secured, reduce the engines and let the tidal effect take the vessel astern, paying out the forward lines as required. When the lines are at the correct length, hold on to let the vessel swing easy between the line of buoys.
3. Make fast forward and stop engines. Send away the stern mooring lines to the aft buoy and make them fast.
4. Slip wires should then be passed to the buoys at each end, once all fast fore and aft.

Securing mooring lines to buoys

Various ports around the world employ different practices for securing the soft eyes of mooring ropes to mooring buoys. The easiest method is to have sufficient numbers of mooring shackles to send away with each eye, which will permit each rope to be separately connected to the ring of the buoy.

An alternative securing method is to use manila lashing tails, of suitable length, to lash in place a toggle passed through the rope's bight and wedged across the rope eye. Where no toggle is available, the sides of the rope eye can be lashed together under the rope's bight.

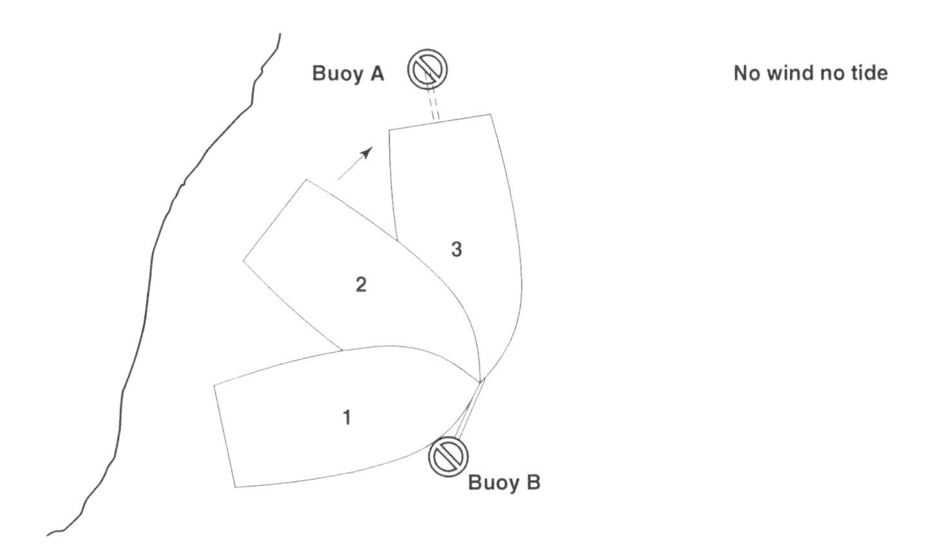

To moor between buoys, to end facing the opposite way.

The vessel should approach 'Buoy B' at a fine angle on the starboard bow and pass a head line, via a mooring boat to be secured to the ring of the buoy. Once the mooring boat is clear, the rudder should be placed hard to starboard with main engine at half ahead. The head line will act as a spring causing the stern of the vessel to be forced upwards towards 'Buoy A'.

Use the same mooring boat to pass a stern line towards 'Buoy A'. Care should be taken not to foul the stern line with the propeller action. Once the line is secured onto the buoy, heave the stern of the vessel into a position of alignment between the two buoys.

Tidal conditions

If the mooring is to be attempted with a right hand or left hand fixed propeller, in tidal conditions the vessel should approach 'Buoy B' as described in the previous example.

Wait for the tide to turn, provide the vessel with a sheer, by use of the rudder and tend the forward mooring as the vessels swings on the tide, to place the stern close to 'Buoy A'. Once in the outward facing position, pass the stern moorings to 'Buoy A'.

Where a ship is equipped with twin propellers or CPP and Bow Thrust (and/or stern thrust), it would not be anticipated that the ship would have to wait for the tides to turn, to complete the mooring.

Departure from Mooring Buoys (Tide Ahead)

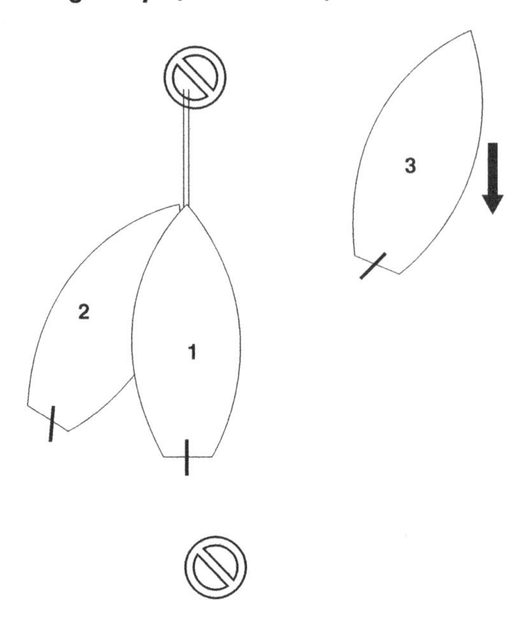

1. Single up to the forward slip wire. As the vessel is held in position with water flowing past the rudder, the ship is still steerable.
2. Helm hard to starboard and the stern could expect to swing outside the line of buoys.
3. Engines ahead, let go the forward line and clear the buoys.

Departing the Buoys (Tide Astern)

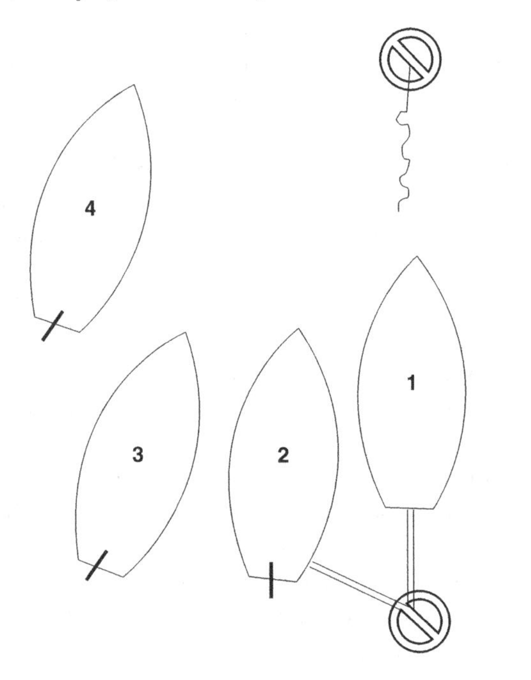

1. Single up to slip wire fore and aft.
2. Slack the forward line. Engines astern, if the stern swings to port with the transverse thrust, then let go the forward line. Allow the vessel to drop astern.
3. Stop engines, let go the aft line. Then engines astern again to clear the buoys.
4. With adequate clearance, stop engines. Engines ahead and apply helm as required.

NB. *Do not attempt to bring the vessel back between the buoys.*

Departing the Buoys (Tide Astern)

When the vessel does not swing with the transverse thrust effect (see previous example).

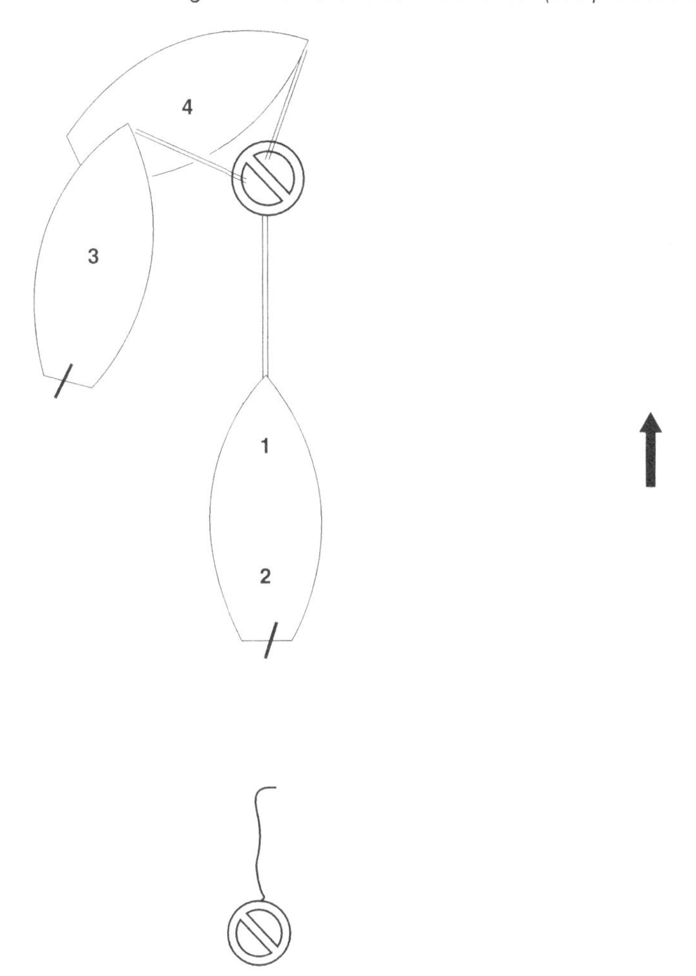

Hanging off a ship's anchor

There may be occasions when it is necessary to hang off the anchor, as when mooring to buoys or in a situation of establishing a 'Composite Towline' (see towing chapter). In the majority of cases and depending on the design of the vessel, the anchor is far too heavy to be cleared and physically hung off by wires. If the operation has to take place at all, it would be carried out with shore side crane facilities or, alternatively, hung off in the hawse pipe by its own anchor securing devices.

If hanging off in the hawse pipe, the bare end of cable would then expect to be deployed via the centre (Bullring) lead (if fitted). Clearly, where a centre lead is not a feature, an alternative mooring strategy would have to be employed.

Previously, vessels with anchors of under 10 tonnes could contemplate hanging the anchor off at the break of the for'cstle head. However, heavy duty wires and associated shackles would have to be available. In the modern day, where ships are carrying anchors in excess of 20 tonnes, the practicalities make such an operation dangerous and foolhardy.

Rigging slip wires (to mooring buoys)

The objective of rigging slip wires to mooring buoys is functional to allow the vessel to control the departure time of the vessel, independent of shore side linesmen.

The wire is rigged as a bight passing through the ring of the mooring buoy, with both parts of the slip secured aboard the mooring deck of the mother ship. Slip wires are generally rigged both fore and aft (ropes are not generally used).

Preparation

1. The wire should be flaked the length of the foredeck.
2. The eye of the wire should be reduced by seizings to allow it to be passed through the ring of the mooring buoy easily.
3. Have a strong messenger line ready flaked to run with the slip wire.
4. Pass both the slip wire and the messenger separately overside towards the water surface, ready to be picked up by a mooring boat.

Operation

The vessel would be expected to be secured by mooring ropes at each end prior to attempting the rigging of slip wires.

a) Pay out one end of the slip wire and one end of the messenger with some slack on each, into the stern part of the mooring boat.
b) Men in the mooring boat would normally coil slack to each line in the sternsheets of the mooring boat to permit flexible operations.
c) The boat would move towards the buoy to permit one man (Buoy jumper) to land on the buoy and pass the reduced eye of the slip wire through the ring of the buoy.
d) Once the eye has been passed through the ring then the messenger can be secured to the eye, to permit pulling the slip wire back to the vessel.
e) As the wire is brought on board the messenger can be detached and both parts of the bight can be turned up on bitts. This will allow either part of the slip wire to be released as and when by the deck crew without calling out shoreside labour.

Precautions

When operating with mooring boats, the personnel in the boat should be wearing lifejackets. If lifejackets are not visible then the mooring officer should offer the use of ships lifejackets for this operation.

When passing the slip wire, the eye should be passed from the upward side of the buoy ring, as this will avoid the eye 'Lassooing' the ring when letting go. When securing the parts of the slip wire, it is recommended that eyes at the ends are not placed over bitts, but 'figure 8s' used for ease of release.

Use of kedge anchor

Kedging is described as the movement of the vessel astern, by means of an anchor or anchors, laid aft of the vessel. Where a ship is equipped with a stern anchor arrangement the employment of its use as a stream anchor, to kedge the ship astern, would not be seen as a difficult operation. Unfortunately, many vessels are not so equipped and if the need to conduct a kedging operation became necessary, limited options may include the use of the ship's spare anchor (if carried).

Spare anchors on board the vessel are meant to substitute for the loss of a main bow anchor, and as such would probably be extremely heavy. To carry out such an anchor astern of the vessel, with the view to kedging the vessel astern, would probably require the use of a tug or similar craft to carry the anchor out and deploy it well aft.

Such an operation would usually be employed in a situation where a vessel has run aground into shallows and is attempting to re-float into deeper water. It must be considered as extenuating circumstances for a ship's Master to even consider trying to kedge the vessel astern. With anchor sizes of large tonnage exceeding 20 tonnes, clearly the practicalities of lifting and deploying such a weight would pose obvious problems.

The use of a ship's lifeboats for carrying out an anchor is not recommended; the sheer weight of a 20 tonnes plus anchor would not give a great deal of safe freeboard to a ship's lifeboat, if so employed, especially if the weight of chain or anchor warp is also included. Such an operation could only be remotely considered with anchors weighing 5 tonnes or less, and even then the operation would be precarious.

Practically speaking, the option to carry out anchors of over 5 tonnes would have to be undertaken by a tug or other similar large craft which could lift and sustain the heavy load. Such a vessel would also require safe means of releasing the anchor when at the position of deployment.

The procedure of 'Kedging' would not generally be undertaken as a stand alone operation. In order to be successful the prudent use of ballast would be employed. Possibly a tug would also be used and the state of the tides with the prevailing weather would warrant consideration.

The foul hawse

The disadvantage of mooring with two anchors is the risk of fouling the anchor cables about each other. This is generally caused by poor watchkeeping practice when a change in the wind direction could result in the vessel swinging in opposition to the lay of the cables.

There are several ways to clear the entwined cables:

a) Use the ship's engines to turn the ship in the opposite direction to the fouled turns at the time of slack water.

b) Engage a tug, to push the vessel around in opposition to the turns, although this method would incur some expense for employing the tug.
c) Use a barge or other similar floating receptacle, break the sleeping cable, and lower the end into the barge. Proceed to pull the barge with the cable, around the riding cable to clear the fouled turns (the ship's boats could provide motive power to manoeuvre the barge).
d) Break the sleeping cable and dip the end about the riding cable, using easing wires from each bow.

It should be noted that to carry out option (d) is a lengthy process, and could only be considered with relatively small anchors. Adequate safe working loads on respective wires would need to be considered with this method. It is also very lengthy, and would have to be managed within the period of the turn of the tide, i.e. 6 hours. This option should be avoided whenever possible, in favour of a more practical alternative.

Attempting to use the engine and turn the vessel about the riding cable (option (a)) would be difficult for a single screw ship but would probably be more viable with twin screws and/or, bow/stern thruster units.

Breaking cables

In any operation where the anchor cable is to be broken, it should be realized that this element alone can expect to take some considerable time. Kenter Joining Shackles are easily separated within the dry dock environment, but when the vessel is operational at sea, the exercise to punch and drift the 'spile pin' can be fraught with difficulties.

Clearing the foul hawse – (breaking and dipping cables)

Procedure

1. Heave up on both anchor cables to bring the foul turns visible above the surface of the water.
2. Lower the ship's boat and lash the two anchor cables together, with a manila rope lashing, in a position below the turns.
3. Pass a preventer wire, on a bight, through the sleeping cable and secure on deck (the preventer is rigged as a safeguard against loss of the chain end, a 24 mm wire minimum, is recommended for use with the average size of anchor).
4. Walk back on the sleeping cable to bring the next joining shackle on deck forward of the windlass.
5. Pass an easing wire from the warping drum to a shackle position on the sleeping cable. Position and rig separate dipping wires from each bow.
6. Break the joining shackle on deck and pay out the easing wire to lower the chain end, clear of the 'hawse pipe'.
7. Pass the dipping wire from the bow in opposition around the riding cable and shackle to the broken chain end. Lead the dipping wire to a second warping drum and heave the wire and the chain end around the riding cable, slacking back on the easing wire.
8. Continue to heave on the dipping wire to bring the end of cable on deck and it will be observed one half turn of the foul has been removed. Detach the easing

wire, leaving it in position in the 'hawse pipe', ready to recover the chain end once all foul turns have been removed.

9. Continue to use alternate dipping (easing) wires to remove each half turn of the foul.
10. With the last half foul turn remaining in the cables, re-secure the easing wire from the hawse pipe and heave the broken chain end up on deck for reconnecting the joining shackle.
11. Once the chain is re-joined clear the preventer.
12. With the aid of a manhelper and knife secured to the end, cut the manila lashing to separate the cables. It would be beneficial to put strain on the lashing by increasing the tension on the anchor cables.
13. Recover the boat and secure the forward mooring deck.

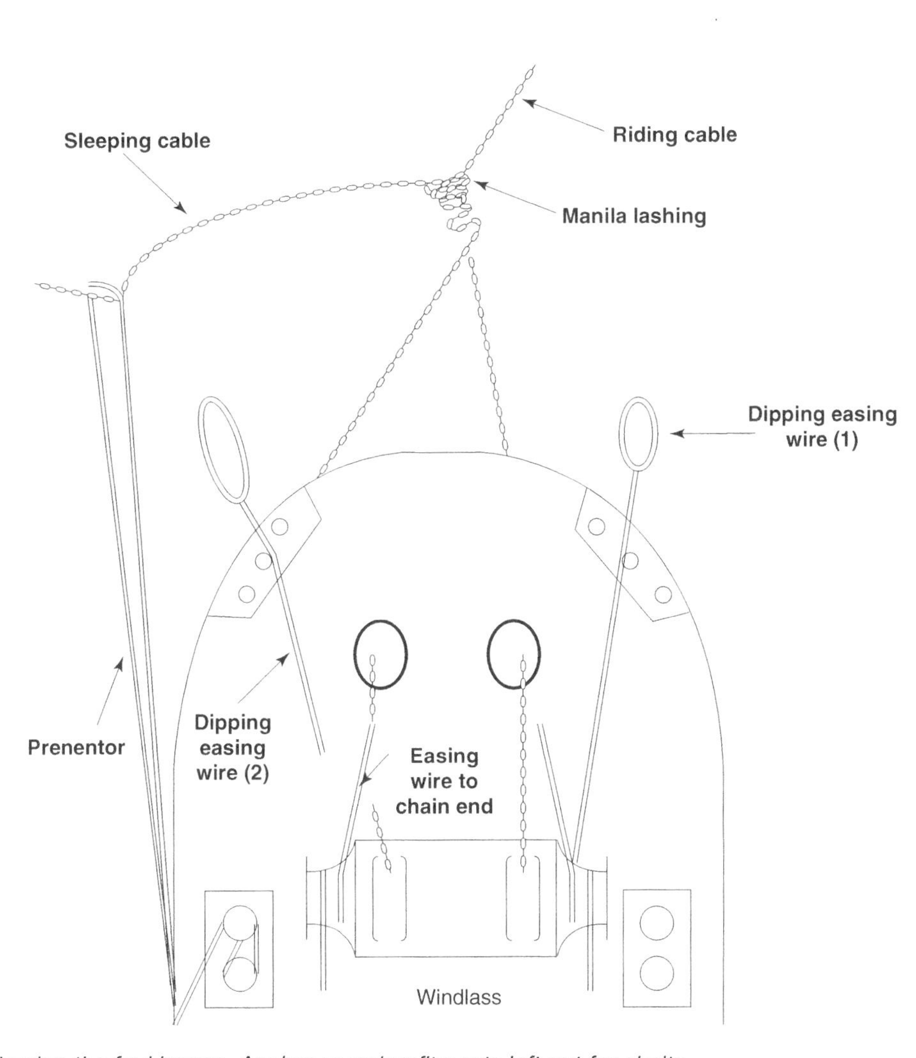

Clearing the foul hawse. Anchor securing fitments left out for clarity.

4 Operations with tugs

Introduction, tug engagement and operations; Bollard pull, towlines care and use, harbour towage; Girting the tug and use of 'Gob Rope'; Towing fitments and towlines, ocean towing, towing methods, composite; Bridle and towing by double units; Towing signals.

Introduction

There are a variety of tug types employed within the marine industry. They include the ocean-going salvage tug down to the smaller harbour traction tug, engaged in and around ports and harbours. Ship handling situations warrant tug use either in a pulling or a pushing mode in numerous situations. The large VLCC or ULCC tankers, for instance, would experience great difficulty in attaining and departing their berths safely, without the assistance of probably at least four tugs.

Entering docks, turning into rivers and engaging in tight manoeuvres, tends to be that much easier and safer with tugs engaged. This is especially so with the large ocean-going vessels that may have limited manoeuvring aids and be restricted to a right hand fixed propeller only.

Marine pilots tend to work closely with tugs and, in the majority of cases, tug masters recognize the authority of the parent vessel and the navigational 'con'. However, tugs can only be considered as being under control when the tug(s) is responding directly and to the desires of that person in command of the operation. Circumstances may make a deviation from the intended movements necessary, but such actions, related to the tug, are not necessarily helpful to the overall operation.

By the very nature of any environment where tugs are engaged, heavy duty operations are envisaged. Towing springs and similar weight bearing ropes are inherently dangerous to personnel who may have to work in close proximity. Full safety aspects should be applied at all times throughout tug operations. Effective communications must be maintained between relevant parties and contingency plans should be in place to reduce the likelihood of accidents throughout this high risk activity.

Harbour Tractor Tugs. An example of a large harbour tug, laying starboard side to the berth in Lisbon, Portugal. Tugs tend to be well fendered, all round, with a bow pudding fender arrangement which permits pushing as well as pulling. A towing hook arrangement is situated just aft of the amidships position, aft of the accommodation block.

Bollard pull

Charter rates for tugs are based on the 'Bollard pull' that the vessel can exert. This is a measure that is determined from trials when the vessel and equipment are new. The towline being secured to a land based anchor point and a load cell measures the strain exerted by the tug when pulling against the fixed point – the higher the bollard pull, the greater the charter rate. Generally speaking, the more powerful the tug, the more the hirer will have to pay. Tugs tend to be hired out for a minimum time period, e.g. 3 hours. The bollard pull is defined by the amount of force expressed in tonnes that a tug could exert under given conditions. A static bollard pull test would be affected by the depth of water, the tugs propulsion rating and the type of propellers fitted.

Safety precautions – handling towlines

Handling and securing of towlines is always a hazardous task and especially so when young or inexperienced seafarers are involved. Everyone concerned on the mooring deck, when tugs are engaged, should be fully aware of the operation and the various stages that the assisting tug is at. Bridge Officers should be aware that from their remote position, they will be relying on two-way communications between mooring stations, and sudden engine movements without adequate warning and timings to deck personnel could have serious consequences.

The following list provides some guidance for deck personnel while Bridge Officers should be thoroughly familiar with the nature of activities when engaging tug assistance.

1. Ship to tug communications should be established well in advance of the designated rendezvous.
2. The designated command authority should be established and each party should be familiar with appropriate manoeuvring signals.
3. Towlines (tug or ship's lines) should be connected by use of good quality heaving lines and/or messengers. The towlines themselves should be of the highest quality.
4. The eyes of towlines should not be placed directly onto 'Bitts', but figure 'eighted' to leave the eye on top. Once secured, the figure '8' wire turns should be lashed to prevent turns springing loose.
5. Personnel should be advised to stand well back from secured towlines. Particularly important once the signal of all fast is made to the tug master. It is normal procedure for a test weight to be taken on the towline, once the line is on the bitts.
6. Personnel should avoid standing in bights or near sharp leads of towlines when the tug is actively engaged in pulling.
7. When letting go tugs, it is normal practice for the tug to manoeuvre to ease the weight on the towline, prior to release of the line. This should be carefully let go, under full control and not just discarded, which could cause injury to persons below on the deck of the tug.
8. Where the ship's towline is used aft and has to be released, an ahead engine movement can usually be beneficial. The wake from the screw race would stream the towline right aft, following release from the tug. This action would expect to provide ample time for the officer on station to land the towline aboard without running the risk of the rope fouling the propellers.

NB. *Towlines carry substantial weight and all personnel, especially young seafarers, should gain experience in their safe handling without being foolhardy. Mooring Deck Officers should actively carry out onboard training in this subject to ensure future safe operations for their personnel.*

Tug towlines

When tugs are engaged, unless designated to push, they will secure with either their own towing spring or a ship's line (usually the best line the vessel has on station). Obvious hazards exist when making lines fast, as in the fact that lines may enter the water in close proximity to turning propellers. To this end, mooring officers are expected to keep the bridge fully informed when running lines overside, especially if and when the ship's engines are still turning propellers. The problem is not as great forward as clearly as it is aft, in a position of the main propulsion. However, bow thrust operation at the wrong moment has also been known to foul ropes.

A towline is passed down from the aft mooring deck to the stern of the tug. The dangers from interaction and the proximity of the two sterns is clearly seen. The manhandling of the tow line is also heavy work and personnel are expected to take all reasonable precautions when securing towlines, prior to a towing operation.

The Panamanian vessel 'Etilico' seen navigating stern first with the tug 'Montsacopa' providing the motive power to a berth in Barcelona, Spain.

Following a turn off the berth the 'Etilico' is seen berthing Port side to the berth in the Spanish Port. The tug's line from the aft position is eased to permit the forward moorings of the ship to be landed. Once the ship is secured alongside, the tug would be dismissed.

Tug operations about the pivot point

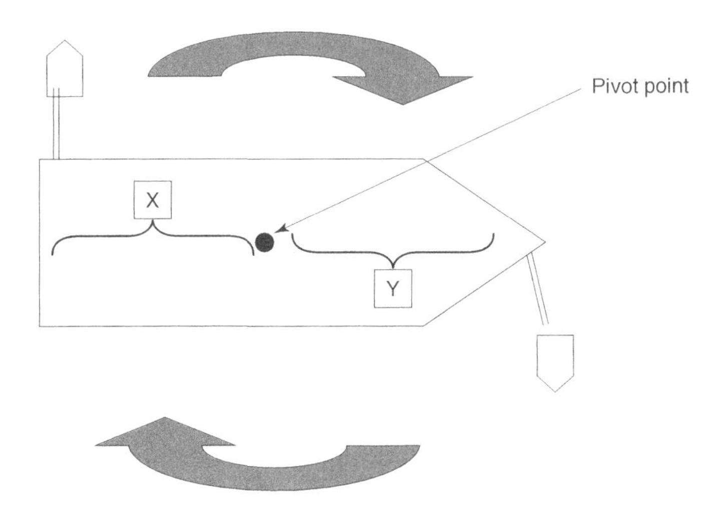

When the vessel is stopped in the water, the pivot point is in approximately the midship's position near to the ship's Centre of Gravity. If tugs are secured at the respective ends fore and aft to turn the vessel, the forces acting on the parent vessel are equivalent to: That force exerted by the aft tug x distance X, while the force exerted by the forward tug x distance Y. These two forces acting about the pivot point generate a couple and cause the vessel to turn about the amidships position; the generated forces being greater than water resistance being experienced by the hull.

***NB**. The diagram shows tugs pulling from their respective positions. It should be realized that if the tugs were pushing with equivalent forces the movement would be the same.*

Where a single tug is engaged – say forward – and the vessel is moving ahead by engines, the pivot point will move to a forward position. The force then exerted to turn the vessel would then equate to that force exerted by the tug x distance Z.

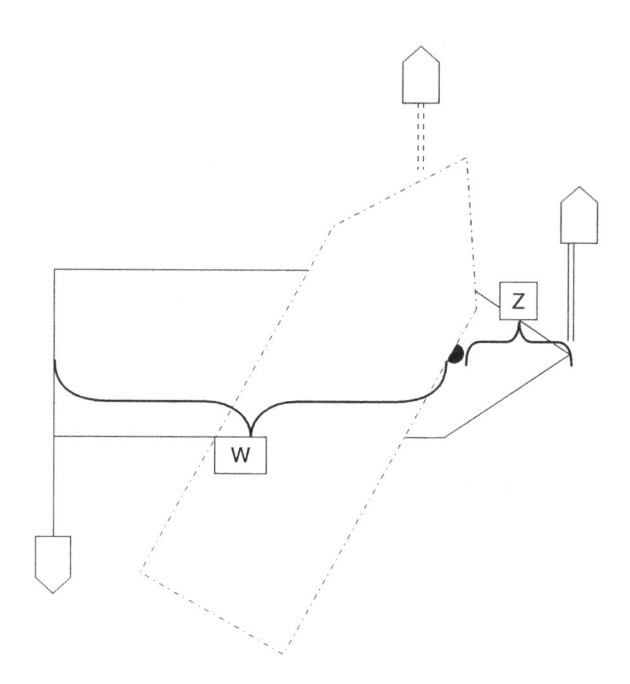

If a tug is engaged aft an opposite force would be exerted equating to: That forces exerted by the aft tug x distance W.

The greater lever W is seen to be far greater than the lever Z. This effectively means that the forward tug would need to exert greater power than the aft tug, in order to achieve the same turning force as the aft tug (if required), at the forward end.

Clearly, the turning capability of an engaged tug on the parent vessel will be directly related to the position of the ship's pivot point and the length of the turning lever (force of the couple) generated.

Multiple tug use

Large vessels like the VLCC and ULCC tankers find it essential to engage several tugs to exercise control of the vessel when in restricted waters as with berthing and unberthing. Tug use is often a combination of pulling and pushing units to assist in turning, steerage and checking adverse external elements like eddies, currents, wind and inactive forces effecting the movement of such massive hulls.

The BP Tanker British Reliance assisted by four tugs. Two bow tugs are engaged in pulling, while two other stern tugs are just visible operating off the port and starboard quarters.

Tugs engaged in pushing

The majority of tugs have the dual capability to engage in pushing as well as in towing a vessel. Pushing has its own merits and is often used to assist large vessels to manoeuvre through dock entrances and narrow access positions. The tug is employed to push the parent vessel off concrete knuckles at the corners of dock entrances.

The tugs which are so engaged are well fendered around the stem and the bow regions to present a soft cushioning pad against the ship's side shell plate. The pushing action is particularly useful when a river tidal stream is causing the parent vessel to set down onto an obstruction.

The 'Jahre Viking' the largest manmade transport in the world (seen prior to being converted into a FSOP) employing a tug off the port bow at the entrance to the Dubia Dry Dock complex.

Girting the tug

Tug Operations – securing the towline. The aft deck of an operational tug engaged in stretching a towline. The line is passed over the towing rails and secured by the 'soft eye' onto the amidships towing hook. A gob wire/rope arrangement is also set up in the slack condition, being led directly from the centre line winch through an aft lead, to a heavy duty shackle over the towline.

Any towing operation is coupled with inherent hazards and therefore carries an element of risk. One of the main concerns by the tug is where the angle of the towline leads towards the beam away from the natural position over the stern. Should the lead of the towline traverse towards the tug's beam it could generate a capsize motion instead of the desired pulling action, the movement is known as 'girting' the tug.

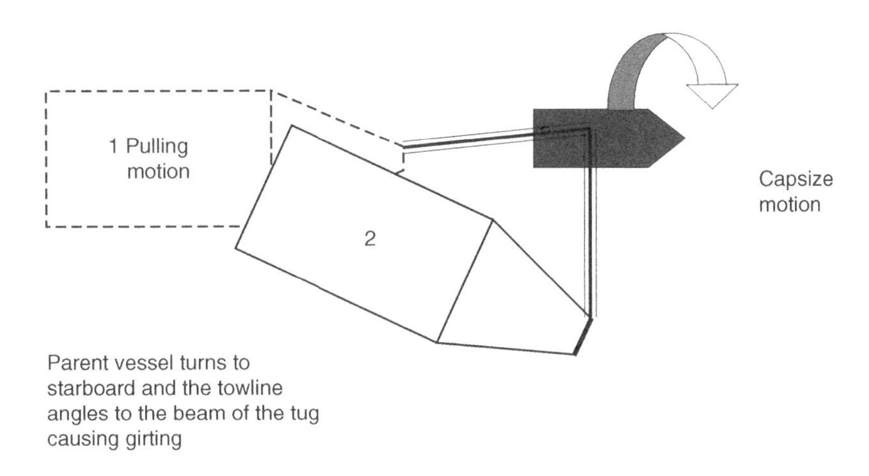

In order to avoid girting the tug, it is normal practice to employ a 'gob rope' (sometimes referred to as a gog rope). It is generally used at the aft end of the tug's deck to change the towing position (fulcrum point) from the amidships position to a position right aft. This effectively generates a turning motion as opposed to a capsize motion, causing a slewing action on the tug. Such a change in motion eliminates the immediate danger of capsize.

The gob rope is rigged in conjunction with a pipe lead or stag horn bitts for the anchored end, with the bight of the rope being led to a winch or capstan. Alternative rigging may employ a wire with a heavy duty shackle about the towline. The wire is led to a winch or capstan at the after end of the tug to draw the towline down to a position away from the amidships position.

Use of the gob rope

Where wire tow ropes are employed and rigged for use in conjunction with a 'Gob Rope' a heavy duty haul wire with a bow shackle is an alternative rig. It is usually secured from a winch drum with the towing wire passing through the shackle at the other end.

Control of wire towing line by means of gob wire and heavy duty shackle.

The gob rope is seen anchored to the 'bitts' and the bight is passed over the towline and led back through the pipe lead onto a capstan. The gob rope can be payed out and slackened off or heaved in close to draw the towline to the aft lead by the capstan operator.

Tugs towing fitments

Tugs towing fitments. The aft towing deck area of the tug 'Montado' of Lisbon, Portugal. The operational area of the tug shows clearly the towing rail across the breadth of the vessel. This is somewhat of a misnomer as the purpose of the rail is not meant to take the weight of the towline, but to permit the towline to ride over it, so protecting the heads of the deck crew situated lower than the height of the tow rail. The eye of the combined rope/wire towing spring is over the amidships towing hook, while the eye and bight are coiled to the starboard side of the centre line capstan.

Sister tugs lie moored alongside each other. The aft decks show centre line 'stag horn' leads for passing towlines onto the towing hooks set aft of the accommodation block. The deck also shows a mooring capstan on the starboard quarter and an electric centre line winch positioned forward of the stag horn.

Tugs and towing operations dictate a variety of towing fitments on many different types of tugs. Many operate with towing hooks or alternatively a towing winch, while many have the flexibility to use both hook and winch from the same operational deck. The leads for the towline to the hook or towing winch will also reflect variety by way of pipe leads or stag horns.

Arguments for and against – towing hooks/towing winches

Seafarers have operated with both towing hooks and towing winches for decades. The simplicity of a hook securing has always been popular as there is seemingly less that can go wrong and the towline can be slipped from the hook very quickly, in the event of emergency. However, where a hook securing is employed, the towline is of a fixed length and generally does not lend itself to easy adjustment.

In comparison with towing hooks, the towing winches have become well-developed and are now widely used. Like any other machinery plant, towing winches must be well-maintained to provide the necessary operational reliability. There is more to go wrong with towlines on winches, but the length of towline can be easily adjusted to either spread the load or ease frictional chafe on bearing surfaces.

Preference for hooks or winches is not usually debated by operational personnel because the engagement will use what is available as and when the tug arrives. The nature of the towing operation will generally lend to the selection of a respective towing vessel with suitable equipment, e.g. docking tug/towing hook, or Salvage tug with towing winch.

Dual purpose tugs are often fitted with both towing hooks and winches and each system will have advantages as well as disadvantages. Reliability and use of each method is derived from experience and practice by operators. They both need degrees of maintenance to be able to continue in safe operations and to this end emergency trips and release gears for winches should be regularly tested.

Towing springs are expected to take exceptional loads and where towlines make contact with bearing surfaces of hooks or winches, it is essential that wires or ropes are not impaired in any way. Correct leads for towlines towards securings should not provide wide or acute angles. Any grooving on leads can cause towlines to jump generating shock stresses to directly affect winch barrels, hook surfaces or even cause the breaking strength of the line to be exceeded.

Towline construction

The construction of wire towlines varies considerably and tends to be dependent on the method and nature of the towing operation. The lighter and smaller towing operation may employ a 6 × 12 wps wire. However, the more popular towlines which generally require greater flexibility employ 6 × 24 wps (FSWR) or the 6 × 37 wps (EFSWR).

Example Sizes	*Minimum Breaking Load*
28 mm (6 × 24 FSWR)	31.3 tons
52 mm (6 × 24 FSWR)	108.0 tons
28 mm (6 × 37 EFSWR)	33.0 tons
52 mm (6 × 37 EFSWR)	114.0 tons

Ocean going – towing operations

Various circumstances may dictate the needs to conduct an Ocean (long distance) towing operation. The vessel intended for tow may be disabled, or she may be on route to be scrapped. Whatever the reason for the towing operation, certain requirements must be complied with to ensure the operation is carried out safely. The following considerations should be evaluated depending on the nature of the operation, assuming that the tow is commencing from a recognized Port of Departure:

1. Is the vessel to be towed in a seaworthy state?
2. If the towed vessel is manned, is the Life Saving Equipment compatible with the number of crew on board?
3. Unless the vessel is being towed as a 'Dead Ship' is sufficient fuel on board?
4. Has a suitable passage plan been adopted for a tug and tow operation?
5. What is the intended speed of the passage?
6. What type of towline is to be used?
7. What length of towline is to be employed and its method of securing?
8. Are the towing arrangements to be inspected by a 'Tow Master'?
9. Has a local and long range weather forecast been obtained for the provisional time of departure?
10. Have the respective Marine Authorities been informed of the operation?
11. Have respective 'Navigation Warnings' for specific areas been posted?
12. Has communication channels between the tug and towed vessel been confirmed? And has communication between the command vessel and the shore side controls been confirmed?
13. Has the command authority between the tug and towed vessel been confirmed?
14. Have contingency plans been considered for: (i) loss or parting of the towline; (ii) poor visibility being encountered on route; (iii) bunker ports or ports of refuge on route; (iv) adequate nautical charts and navigation equipment being available; (v) special signals in the event of radio failure?

Own vessel towing operations

On occasion it may arise that a Master's own vessel may encounter circumstances where it is a requirement to participate in carrying out a towing operation. In such circumstances consideration should be given to the following:

1. Has the ship owner's permission been given to carry out the towing operation?
2. Will the Charter Party and respective Clauses or the Charter's Agreement allow a towing operation to be conducted?
3. Will the vessel have enough fuel to carry out the towing operation?
4. If taking up the tow are there any cargo ramifications for own vessel, i.e. perishable cargo being carried?
5. Are the deck fittings capable of accepting the capacity of a towing operation?
6. Is the value of the towed vessel and its cargo worth the effort?
7. Is the main engine power of own vessel adequate to handle the tow operation?
8. Has own ship an adequate towline of sufficient strength and size and length?

9. Under what agreement will the towing operation be conducted, i.e. Contract of Tow, or Lloyds Open Form?
10. If the tow is engaged, can own ship still reach its own loading port without incurring penalties?
11. Is the vessel capable of completing the proposed passage plan safely?
12. Has the vessel obtained respective clearances and posted navigation warnings?
13. Is the towed vessel manned and in a safe condition to be towed?
14. Have contingencies for the proposed tow operation been considered?

Care of the towline

In any towing operation the essential element is the towline. Its selection in the first place should take account of strength. Its length and size will reflect the elasticity but will also influence the handling position of the vessel being towed. A short length of towline is easier to control and reduces 'Yaw' on the vessel being towed, whereas a long length in the towline has greater shock absorption throughout its length.

Good leads must be provided for every towline and should favour less friction bearing surfaces where possible. Sharp angled leads should be avoided at all costs. Adequate lubricant should be applied regularly to the bearing surface of leads to reduce friction burns. The towline should also be able to be length adjusted, to ensure even wear and tear on a variable length of the towline.

The speed of operation should consider the tension in the towline and not be such as to cause the line to snatch. Regular checks on weather forecasts should allow the line to be adjusted in ample time, prior to entering heavy weather. In the event that the towline is parted, suitable means of recovery should be kept readily available throughout the period of tow.

The long and short towline

A short towline is much more liable to 'snatch' and part than a long one, although a short length is easier to control and steer with than a long towline. Long towlines lend to excessive yawing and make the operation difficult to steer.

Long towlines also have greater scope for adjusting the length but are difficult to manage recovery in the event that the line parts under tension. Towlines and the miscellaneous factors affecting a staged towing operation would be inspected by a Tow Master. Advice from the Tow Master regarding the length and type of towline should be followed or the ship's insurance may become invalid.

Towline safety

The area of the towline and its securing should be cordoned off and unauthorized persons should be denied access to the securing. Regular inspections should be made to the bearing surface of the towline and a watchman duty would be scheduled.

Course alterations should be staged to avoid unnecessary tension to the towline and not cause the lead to be become excessively angled. Anchor(s), at least one, should be kept readily available for emergency use when in shallow waters.

The passage plan should highlight navigation hazards and effective communications should be in place between the towing vessel and the vessel being towed throughout the operation. Correct signals, lights and shapes should be displayed during the daylight and night time passages as per the Regulations for the Prevention of Collision at Sea.

The composite towline

From the disabled ship's point of view (when requiring a tow), the best towline arrangement would probably be by use of a towing bridle. However, such an arrangement would clearly not be possible to establish in open waters, with variable weather conditions, even if the ship had the resources to set up a bridle. The most practical accomplishment would more likely be a composite towline which would be established by the ship's anchor cable being joined to the tug's towing spring.

The composite towline would employ one anchor cable, leaving the other anchor available for use in an emergency. The advantages of the composite towline is the lead by way of the hawse pipe, which is already available, and the length of the anchor cable to the towline would be easy to adjust via the ship's windlass.

The weather conditions would dictate whether the anchor would be left *in situ* on the end of the cable or whether the anchor would be hung off in the hawse pipe, with the cable being broken at the ganger length and the bare end passed through the centre lead.

NB. Hanging off an anchor at the shoulder, in open waters, with such heavy anchors in current use, is considered virtually impractical.

Leaving the anchor on the cable would be labour saving, and relatively risk free. While, at the same time, the anchor left on the stretched composite towline could expect to act as a damping effect on any towline movement. Such movement is obviously influenced by the prevailing weather/sea conditions and the length of the towline. A long length in the towline generally reduces the snatch effects but sometimes makes it more difficult to steer without incurring excessive 'yawing'. Whereas a short towline is better to control the vessel being towed, but lends to snatching over a short length which may cause the towline to part.

The weakest part in a composite towline operation is always the eye of the towing spring where it is joined to the ship's anchor cable. If the tow is going to part it will probably be at this point. In confined waters the towline may have to be shortened and where this is the case, a reduction in towing speed may compensate for increased stress being incurred. Alterations of course should be carried out in small stages 20° or less and large alterations should be avoided at all times. Speed reductions should also be gradual to avoid the towed vessel making contact with the towing vessel.

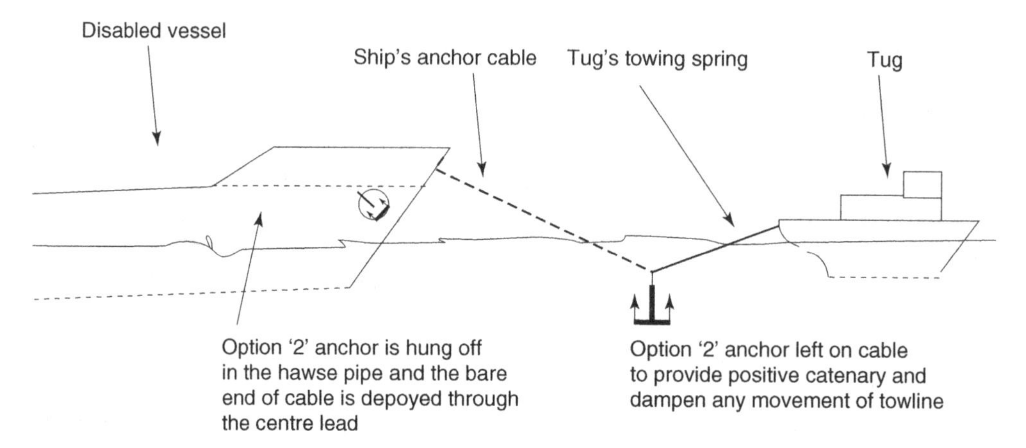

Length of towline can be adjusted easily by the disabled vessel, if manned. Second anchor of disabled vessel is left ready for emergency use.

Statutory towing requirements for tankers

In 1983 IMO made it compulsory for new tankers over 20,000 grt to be fitted with emergency towing fitments. Subsequent amendments make it a requirement that all tankers over 20,000 grt must be fitted with emergency towing fitments fore and aft.

Such fittings usually consist of an anchor point in the form of a 'Smit Bracket' to which is secured a chafing chain led through a ship's lead. The purpose of this arrangement is to permit a tug to take the vessel in tow in the event that the vessel starts to founder. The principal idea being that the ship's crew, prior to abandoning the vessel, will have the opportunity to deploy the chafing chains through the leads, to be accessible overside.

Even with no persons left aboard, weather permitting, a salvage tug would then have the ability to secure a towline to the anchored chafe chain.

Different arrangements are current within the industry and some are fitted with a towline which can be quickly secured to the chafing chain. Clearly, the carriage of the vessel's own towing spring would provide a certain amount of flexibility, which would permit any vessel to take up the tow, not just a designated tug.

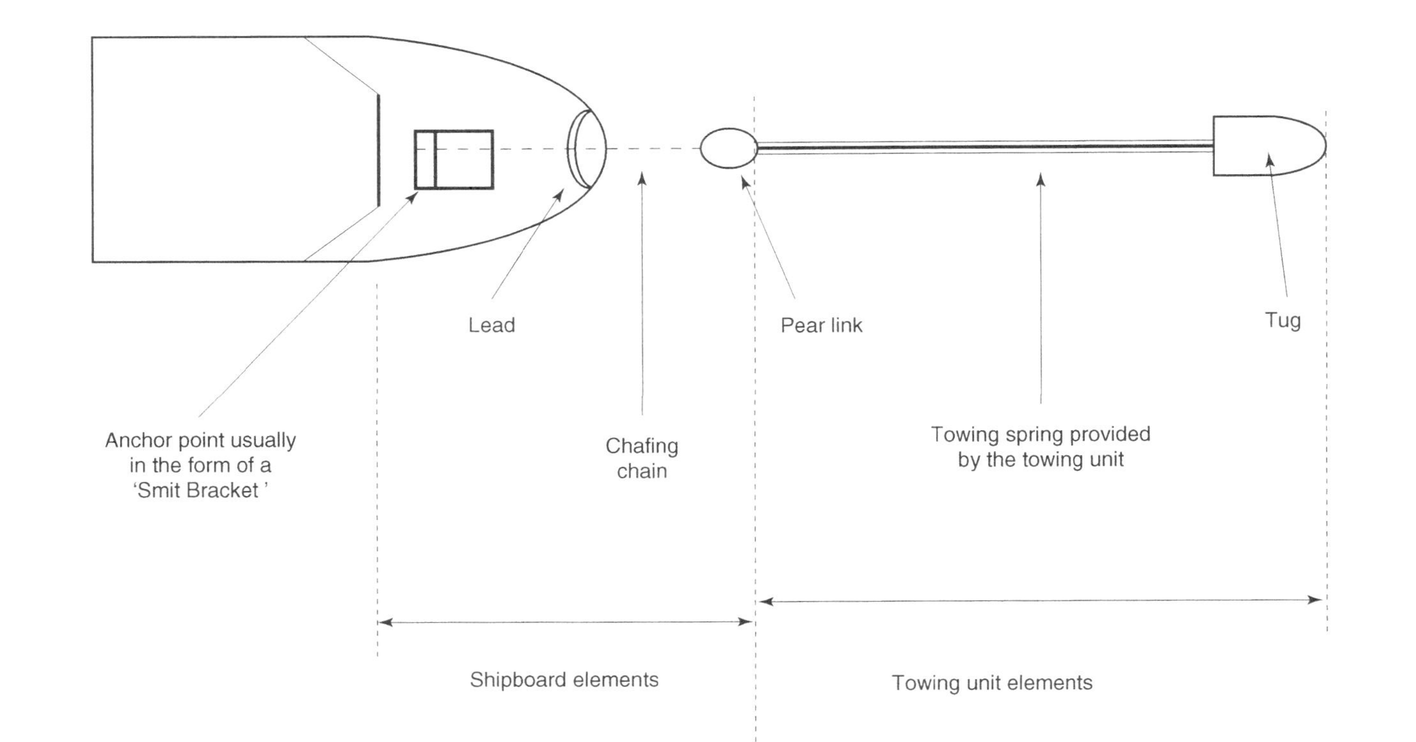

Statutory emergency towing arrangements for larger tanker vessels. The shipboard elements of the emergency towing arrangement include the anchor point securing, a lead to accept the chafing chain and the pear link. The chafe chain is generally stowed in a box arrangement to one side of the forecastle head from where it can be easily deployed. Alternatively, some vessels may carry their own towing spring unit below decks which can be easily shackled to the chafe chain and deployed through the purpose designed lead.

The towing bridle

This method of towing employs a chain bridle secured to the vessel being towed. The bridle is then connected to the towing spring of the towing vessel. Unfortunately, few vessels carry the means to construct an effective bridle arrangement and even if they did, crews would find the task very heavy work indeed.

The chain bridle is more commonly established by a rigging gang when the vessel is in port. Cranage is usually available when in port and the arrangement can be more easily secured. The towline and the bridle would normally be inspected by a Tow Master usually appointed by the Classification Society on behalf of the Underwriters. One of the tasks would be to ensure that the bridle and the towline are 'sound' and not likely to part when engaged in the towing operation. The Tow Master would also expect to be satisfied that a contingency plan and an alternative towing method are available in the event of the towline parting.

The fact that a bridle is being employed would infer that any weak point in the towing arrangement is in the towline itself and not in the bridle fitting. Clearly, if the towline did part on passage, the task of re-securing to the bridle would mainly fall to the tug's crew and not the seaman of the vessel being towed.

The size of chain used in a bridle will vary in comparison to the size of the vessel. However, handling chain cable is never easy and must always be considered as heavy duty labour. A risk assessment prior to commencing such work would be considered an essential element of safe practice.

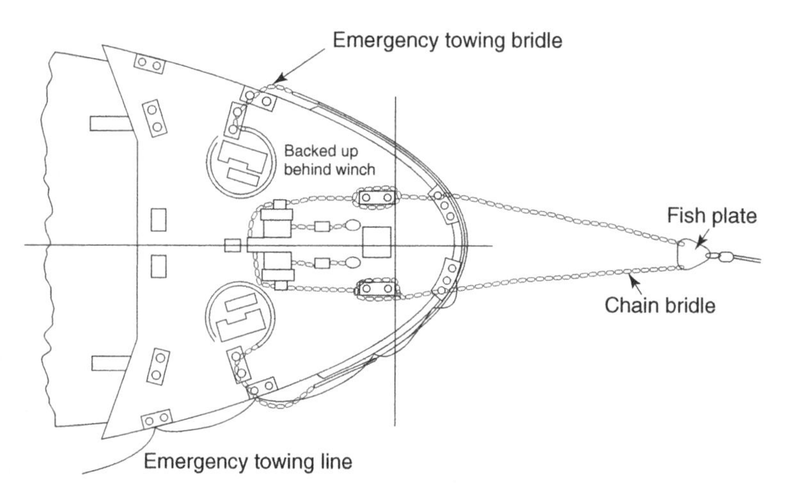

Purpose-built chain bridle rigged from a conventional forward mooring deck. The emergency towing bridle is also established as a precautionary, contingency measure.

Ocean towing with two or more tugs

Where heavy towing operations are undertaken, the use of multiple tugs is not unusual. This is especially so around the offshore industry where installations or platforms are established, prior to becoming 'Hot'. Where double units are engaged on a towing operation, it is normal practice that respective towlines are secured at alternative lengths to provide collision avoidance.

Where a third towing unit is employed, this may be engaged astern of the towed object in order to provide steerage and guidance to the leading tugs. Alternatively, a third tug could be engaged as a centre lead towing unit, accompanied by sister tugs angled off to either side. In such a configuration all three tugs would provide a uniform pull to generate forward motion on the unit or object being towed.

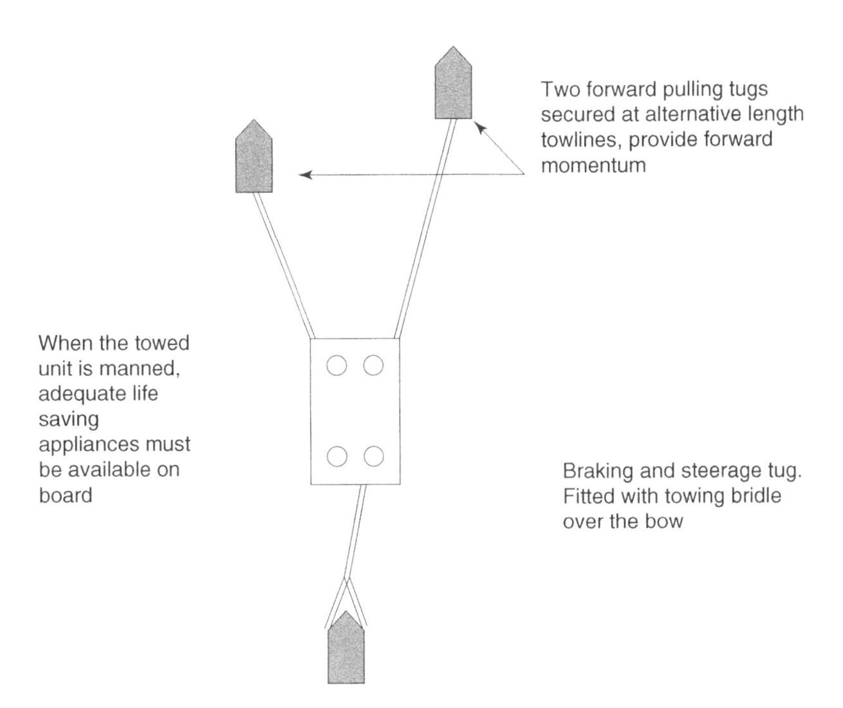

Towing signals

When vessels are engaged in towing, depending on their status will dictate the variety of signals that they must display. These towing configurations are further complicated when towing either another vessel or when towing a partially submerged object. The following configurations provide a guide to the various towing displays and the correct use of respective signals.

A vessel engaged in towing, where the tug is probably more than 50 metres in length and the length of tow is more than 200 metres from the stern of the tug to the stern of the last vessel towed

The day display signal carried by the tug and the vessel towed will be a Black Diamond shape, where it can best be seen on each vessel

Additionally where the tug is experiencing difficulty with the towing operation, she may also show by day restricted in ability to manoeuvre signals, namely by day Black Ball, Black Diamond, Black Ball

By night:
Red, white, red lights in a vertical line

Where the tug is less than 50 metres in length, but the tow arrangement is more than 200 metres then the configuration of signals is as below.

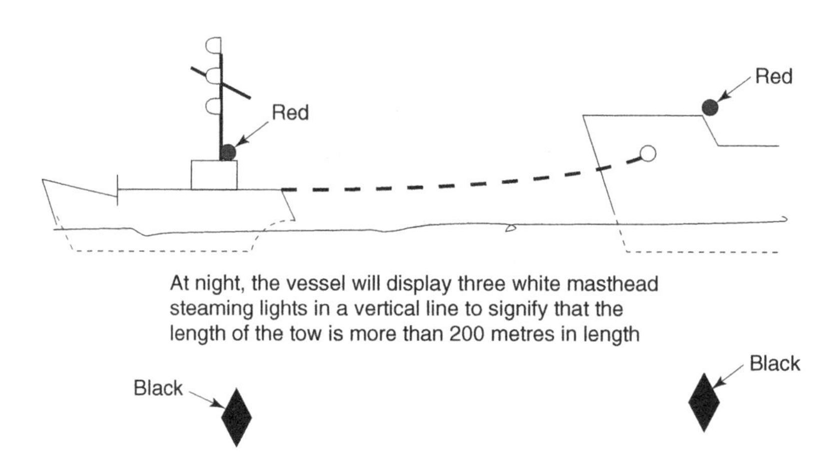

● ♦ ● By day, each vessel will display a black diamond shape where it can best be seen. The restricted inability to manoeuvre signals would also be displayed if the towing vessel was experiencing difficulty managing the tow

Where a vessel is engaged in towing where the length of the tow is not more than 200 metres from the stern of the tug to the stern of the last vessel towed, no day signal will be displayed by either vessel.

***Exception**: Where the towing vessel is experiencing difficulty with the towing operation, even though the length of the tow is short, she may still exhibit the restricted in ability to manoeuvre signals by day and by night.*

Towing light

At night, the power-driven vessel engaged in towing must also display a yellow towing light in a vertical line above its own stern light. The characteristics of this light are to be of the same nature as the stern light, except being 'yellow' in colour; the purpose of the towing light being to warn other vessels approaching from astern, of the nature of the operation and also to assist the steerage, by the towed vessel.

When towing inconspicuous partially submerged objects

When engaged in towing a partially submerged object or combination of vessels/ objects, the towing vessel will carry the same signals as for towing an ordinary vessel.

By day: The unit being towed will always show a Black Diamond shape aft, regardless of the length of the tow operation. In the event that the length of the tow is greater than 200 m in length, then the unit would have an additional black diamond shape displayed forward.
By night: The partially submerged object would carry an all round white light at or near the forward end and an all round white light at or near the after end. In the event that the length of the object exceeds 100 m then an additional all round white light would be exhibited between the fore and aft all round white lights.

Where the object is greater than 25 m in breadth, two additional all round white lights would be carried at or near the extremities of the breadth.

NB. Dracones need not exhibit the light at the forward extremity.

5 Emergency ship manoeuvres

Introduction; Heavy weather operations; Synchronized rolling and pitching; Search and rescue search options, manoeuvring for man overboard; Search patterns, choice and aspects; Holding off a lee shore; Use of sea anchors; Maritime assistance organizations; Collision and actions following grounding, beaching, pollution; Damage control; Manoeuvring in ice; Emergency steering operations.

Introduction

Shipping the world over is notorious for experiencing the unusual and the unexpected. In most cases if and when routine practice goes wrong, the weather is usually a key element which influences the cause and very often the outcome. The other variable is often the human element which can work for, or against, the well being of the ship.

The loss of engines when off the lee shore is the classic nightmare of any ship's Master. Equally, the steering control of a ship could be lost. In either case, the root cause may often be traced back to wear and tear or lack of affective maintenance. When coupled with heavy weather, it can easily run to a comedy of errors, ending in a total constructive loss.

Good seamanship to one man is perceived differently by another. Improvisation, with a 'jury rudder' could well save the day, but the use of a high-powered tug could be a more viable and confident alternative to take the vessel out of danger. The marine environment has never been in a situation to be able to dial the emergency services. Ships have had to sustain themselves in all manner of emergencies close to or far out from the nearest land.

The man overboard, the grounding, or the collision, could all require the expertise of emergency ship handling procedures to sustain life and protect the environment. Such incidents need to be tackled with experience, seamanship-like practice and, very often, with an ample portion of common sense. The experienced ship handler has skills considered essential in many emergency situations, even if it is only turning the stern to the wind with a fire on board.

With some forethought it is clear that most incidents can be accommodated with pre-planning, and it is the function of this chapter to highlight typical incidents that may be useful within emergency plans and checklists. An active response will often require the combined skills of on-board personnel, engineers, fire fighters and the ship's handler as a typical example; the outcome being directly related to the safety of life at sea and the best manner in which to provide a protective shield.

When things do go wrong the ramifications to passengers, crew and the environment can be catastrophic. It is at these times that the real art of seamanship must come to the fore and hopefully the correct action would lead to recouping any adverse situation.

Heavy weather precautions (general cargo vessel) open water conditions

Stability

Improve the 'GM' of the vessel (if appropriate)
Remove free surface elements if possible
Ballast the vessel down
Pump out any swimming pool
Inspect and check the freeboard deck seal
Close all watertight doors.

Navigation

Consider re-routing
Verify the vessel's position
Update weather reports
Plot storm position on a regular basis
Engage manual steering in ample time
Reduce speed if required and revise ETA
Secure the bridge against heavy rolling.

Deck

Ensure life lines are rigged to give access fore and aft
Tighten all cargo lashings, especially deck cargo securings
Close up ventilation as necessary
Check the securings on:

Accommodation Ladder
Survival Craft
Anchors
Derricks/Cranes
Hatches

Reduce manpower on deck and commence heavy weather work routine
Close up all weather deck doors
Clear decks of all surplus gear
Slack off whistle and signal halyards
Warn all heads of departments of impending heavy weather
Note preparations in the deck logbook.

NB*. When a ship has a large GM she will have a tendency to roll quickly and possibly violently (stiff ship). Raise 'G' to reduce GM. When the ship has a small GM she will be easier to incline and not easily returned to the initial position (tender ship). Increase GM by lowering 'G'. Ideally, the ship should be kept not too tender and not too stiff.*

The Masters/Chief Officers of vessels other than cargo ships should take account of their cargo, e.g. containers, oil, bulk products, etc., and act accordingly to keep their vessels secure. Long vessels, like the large ore carriers or the VLCC, can expect torsional stresses through their length in addition to bending and shear force stresses.

Re-routing to avoid heavy weather should always be the preferred option whenever possible. If unavoidable, reduce speed in ample time to prevent pounding and structural damage to the vessel.

Bad weather conditions – vessel in port

The possibility of a vessel being in port, working cargo, and being threatened by incoming bad weather is of concern to every ship's Masters. Where the weather conditions are of storm force as, say, with a tropical revolving storm (TRS), it would be prudent for a vessel to stop cargo operations, re-secure any remaining cargo parcels, and run for open water. Remaining alongside would leave the vessel vulnerable to quay damage. Provided the weather deck could be secured, the vessel would invariably fare better in open waters than in the restricted waters of an enclosed harbour.

In the event that the vessel cannot, for one reason or another, make the open sea, the vessel should be either moved to a 'Storm Anchorage', if available, or well-secured alongside. It is pointed out that neither of these options is considered better than running for open waters.

Storm anchorage – if the ship is well sheltered from prevailing weather and has good holding ground, this may be a practical consideration with two anchors deployed and main engines retained on stand-by.

Remaining alongside – increase all moorings fore and aft to maximum availability. Lift gangway, and move shore side cranes away from positions overhanging the vessel. Carry out and lay anchors with a good scope on each cable, if tugs are available to assist. Ensure that engines and crew are on full stand-by, for the period when the storm affects the ship's position.

In every case, cargo and weather decks should be secured and the vessel's stability should be re-assessed to provide a positive GM. Free surface effects should be eliminated where ever possible. Statements of deck preparations should be entered in the logbook, weather reports should be monitored continuously and the shore side authorities should be informed of the ship's intentions.

NB. Where the intention is to run for open waters, the decision should be made sooner rather than later; for a vessel to be caught in the narrows or similar channel by the oncoming storm, could prove to be a disastrous delay.

Abnormal waves

The sea area off South Africa experiences abnormally large wave activity, and the shipping industry generally has been well aware of these conditions. However, more recent research from satellite imagery has shown that abnormal waves are not restricted to just this area, but can be experienced virtually anywhere in the world's oceans.

These large waves, if encountered – especially by the longer and larger vessels like the VLCC or the long bulk carrier – pose a great threat. In the situation where a ship breaks the crest of such a wave, the danger experienced has been described as looking down into a 'hole in the sea'. Violent movement of the vessel into the trough could expect to generate, at the very least, structural damage; while the worst case scenario might be that the ship's momentum in the downward direction is so steep that the ship lacks the power to recoup, to ride the next wave.

Good 'Passage Planning' to avoid areas with a reputation of abnormal waves is clearly a prudent action. While a reduction of speed in heavy weather is considered as general practice, may go some way to combat the effects of that rogue wave, if encountered unexpectedly.

Synchronized rolling and pitching

Rolling

Synchronized rolling is the reaction of the vessel at the surface interacting with the 'period of encounter' of the wave. This is to say that the period of the ship's roll is matching the time period when the wave is passing over a fixed point (the position of the ship being at this fixed point). The clear danger here is that the ship's roll angle will increase with each wave, generating a possible capsize of the vessel. The period of encounter and the increasing roll angle can be destroyed by altering the ship's course – smartly.

This scenario is always caused by 'beam seas' generating the roll and the Officer of the Watch would be expected to be mindful of any indication of the vessel adopting a synchronized motion. The Officer of the Watch would react by altering the course and informing the Master, even if the condition is only suspected.

Pitching

This condition is again caused by the ship interacting with the surface wave motion but when the direction of the 'sea' is ahead; the movement of the vessel being to 'pitch' through its length, when in head seas. The danger here is that the period of wave encounter matches the pitch movement and the angle of pitch is progressively increased. Such a condition could generate violent movement in the fore and aft direction, causing the bows to become deeply embedded into head seas.

The condition can be eliminated by adjusting the speed (reducing rpm) to change the period of wave encounter. It is not recommended to increase speed as this could generate another condition known as 'pounding'. This is where the bow and forward section are caused to slam into the surface of the sea, such motion causing excessive vibration and shudder motions throughout the ship. This latter condition can cause structural damage as well as domestic damage to the well being of the vessel.

Pooping

A condition which occurs with a following sea when the surface wave motion is generally moving faster than the vessel and in the same direction. The action of pooping takes place when a wave from astern lands heavily on the after deck (poop deck). The size of the wave, if large, may expect to cause major structural damage and/or flooding to the ship's aft part.

With the direction of the sea from astern, some pitching motion on the vessel can be expected and the following sea generally makes the vessel difficult to steer, with the stern section experiencing some oscillations either side of the track.

Ship movements in fire emergencies

Emergency 1. Fire at sea

The discovery of a fire at sea can expect to be rapidly followed by the sounding of the fire alarm. This would alert personnel to move towards their respective fire stations, inclusive of the Navigation Bridge and the Engine Room.

The Master would essentially take the 'conn' of the vessel and place the engines on stand-by manoeuvring speed. Ideally, the ship's position will be charted and in the majority of cases it must be anticipated that the vessel's course will be altered to one that will put the vessel stern to the wind (adequate sea-room prevailing). This action combined with a speed adjustment being designed to reduce the oxygen content into the ship and provide a reduced forced draft effect that would probably occur, if the vessel continued into the wind.

The situation of altering course to place the wind astern is not always beneficial, especially if the fire is generating vast volumes of smoke. Such a situation may make it prudent to take a heading that the forced draft from the wind would clear smoke away from the vessel and permit improved fire fighting conditions to prevail.

Each scenario will be influenced by various factors, not least the nature of the fire, and what is actually burning. In the event of an engine room fire, where total flood CO_2 is employed, then the ship will immediately become a 'dead ship'. Such a situation would invariably leave the vessel at the mercy of the weather conditions. This situation may dictate the need to engage with an ocean-going tug at a later time, once the fire is out.

A ship's cargo hold fire will have alternative criteria, depending on the nature of the cargo. An example of this can be highlighted with a coal fire, where the course of the vessel is altered to seek a 'Port of Refuge' in the majority of cases. Circumstances in every case will vary and reflect the ship's movements. Influencing factors throughout an incident will most certainly be the weather conditions prevailing at the time, the geography of the situation and whether a ship's power can be retained, albeit to a reduced degree.

Masters may consider taking the ship to an appropriate anchorage if available, with the view to tackling the fire at a reduced sea-going operational level, so to speak. Also, the availability of shore side assistance by launch or by helicopter tends to become viable off a coastal region as opposed to a deep sea position. The advantage of this option is that specialized fire fighting equipment, supplies and manpower can generally be made more readily available.

Finally, it becomes a Master's decision at what time the fire is declared out of control and that the vessel must be abandoned. Such a decision is not taken lightly, knowing that the vessel provides all the life support needs for passengers and crew. Taking to lifeboats in open sea conditions might present another set of problems, and possibly becoming even more subject to weather conditions.

Emergency 2. Fire in port

In every case of fire, whether it is of a minor or major incident, the person so discovering the outbreak should immediately raise the fire alarm, without exception. In the case of the fire in port, the raising of the alarm should also incorporate the calling of the 'Local Fire Brigade'. This action is probably best achieved by means of the ship's VHF via the Port Authority, allowing the brigade to start to move sooner rather than later.

Aboard a vessel in port working cargo, then, it would be anticipated that all cargo operations are ceased and that non-essential personnel, i.e. Stevedores, are ordered ashore. The purpose of removing non-essential personnel from the vessel is primarily to reduce the potential loss of life.

Any discovery of fire and the subsequent sounding of the alarm system could expect to generate positive activity amongst the crew on board the vessel. It must also be appreciated that essential members of the ship's fire fighting teams may be ashore at the time of the outbreak. This would clearly leave fire parties deficient of key personnel. Such a situation could be immediately rectified if the vessel had previously conducted drills, in which 'job sharing' was common practice on positive action drill type activities. In any event the crew would be expected to tackle the fire immediately, even as a holding operation until the fire brigade arrived.

Drill duties and fire fighting activities could expect to cover the following:

1. Manning of the bridge and the monitoring of communication systems.
2. Chief Officer's messenger being established at the head of the gangway to make contact with the incoming 'Fire brigade' personnel.
3. Establishing hose parties and damage control parties (boundary cooling around the six sides of the fire being a positive start). While damage control parties could expect to isolate the fire area by closing down all ventilation in the vicinity.
4. Where direct contact is to be made with flame or smoke, then Breathing Apparatus parties would need to be established in order to provide some containment of the outbreak.
5. Chief Officers would be expected to supply the Fire Brigade with the following:
   a) The cargo plan or general arrangement plan of the effected and adjacent areas of the fire region. Stability information and relevant cargo details and a known list of persons onboard and/or missing.
   b) Place all available crew members on an alert status and engine room personnel on stand-by inside the engine room.
   c) Provide the Fire Brigade with the 'International Shore Connection'.

Tanker vessels – on fire in port

Additionally:

The ship's moorings would be tended and attention paid to the use of fire wires in the fore and aft positions. Tugs may be called to tow the vessel from the berth to reduce immediate danger to the terminal. In such an event, shore side moorings would need to be slipped or cut, and the gangway stowed or sacrificed.

Tugs working around oil terminals are usually equipped with water/foam monitors and these may be brought to bear as the vessel is cleared from the berth. It must be appreciated that the ship's engines may have been shut down while in port, and these may take some time to warm through and become operational. In any event, the use of engines as soon as possible would be considered an essential element of manoeuvring the ship into safer waters as soon as practical after the outbreak.

Communications to tug masters and shore side authorities will also play a major role in achieving a successful resolution of the situation. The weather conditions will also influence the outcome and progress of fire fighting operations.

The 'Pacific Retriever', an anchor handling vessel, displays its powerful fire water monitors while on trials off the Coast of Korea.

Search and rescue manoeuvres

Ships with their Masters and crews can expect to be called to respond to a variety of maritime emergencies. The handling of the vessel in a man overboard situation, for example, could involve one of several types of manoeuvres, depending on prevailing circumstances. Escalation from the immediate incident can progress rapidly when distressed person(s) are not located and recovered quickly.

The apparent loss of a man, or a transport unit, can expect to generate a variety of 'Search Patterns' conducted with one or more units being involved. Reference to the IAMSAR volumes provides an immediate direction for Masters of search units so engaged. However, experience of search procedures must be considered an essential element towards attaining a successful outcome.

Many operations these days are involved more and more with aircraft assistance of the fixed winged variety or rotary blade helicopters. Their height and speed tend to make them ideal for location, although their payload and ability to recover is often not always a practical proposition. The need for a surface craft recovery, especially for large numbers, becomes the only way to gain recovery from the water.

In all of this, the ship handling skills and the control exercised by ships' Masters is seen as being a highly valued element to any operation. The incident with the 'Ocean Ranger' offshore installation in 1982 (the installation capsized when hit by a huge wave) saw so-called survivors only twenty metres away from rescue, only to have the survival craft capsize, with the loss of all its occupants (in all, 84 lives were claimed during that incident). The old adage, that you are not a survivor until landed in a safe haven, tends to bear some reality to common sense.

Masters who find themselves involved in search and rescue operations will usually be co-ordinating their vessel's movements with an On Scene Commander (OSC) or an On Scene Co-ordinator linked directly to a marine rescue centre, ashore (MRCC).

The ship handling aspect of such an operation will be accompanied heavily by internal and external communications. Support internally from a bridge team will be coupled with external support from several outside agencies such as: Meteorological Authorities, Ship Reporting Agencies, Military units, Coast Guard Organizations and not least, other shipping traffic in the vicinity.

The outcome to an incident will generally involve an element of luck but clearly experience, education, modern facilities, information technology, etc. can move an operation along that much quicker, and with more effect. This is especially so when vessels are fitted with enhanced manoeuvring aids, twin/triple/quadruple screws, bow and stern thrusters, stabilizers, etc. and backed by powerful main engines.

Manoeuvring for man overboard

In every man overboard incident it would be expected that the Officer of the Watch would carry out the following simultaneous actions:

1. Place the ship's engines on 'Stand-By'.
2. Release the bridge wing lifebuoy.
3. Raise the general emergency alarm.
4. Adjust (or be ready to adjust) the helm to manoeuvre the vessel.

It should be realized that subsequent additional actions will be required after the immediate, four recommended actions. At the same time, it should also be realized that stopping the vessel and having a 'dead ship' will not help the man in the water or the recovery situation.

The Williamson Turn – man overboard manoeuvre

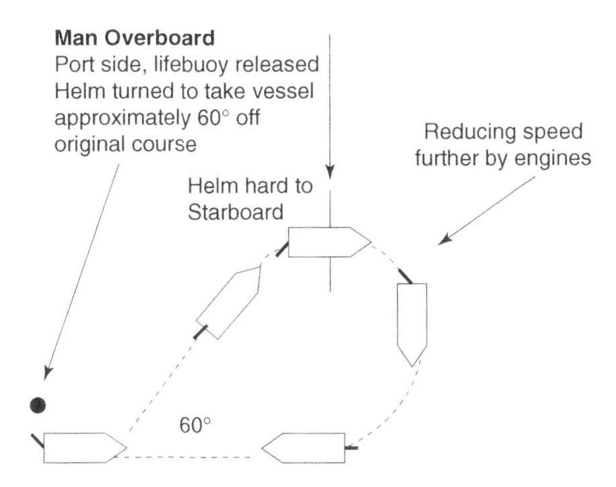

The Williamson Turn. A vessel on reciprocal course, search speed approximately 3 knots. Position the vessel to the weather side of the person in the water.

On approach to the man overboard position, the Chief Officer would be ordered to turn out the rescue boat (weather permitting) and prepare for immediate launch with the boat's crew wearing lifejackets and immersion suits. The ship's hospital would be ordered onto an alert status and be ready to treat for shock and hypothermia. A vessel so engaged would expect to have full communications available throughout such a manoeuvre.

The Delayed Turn – alternative turning manoeuvre for man overboard

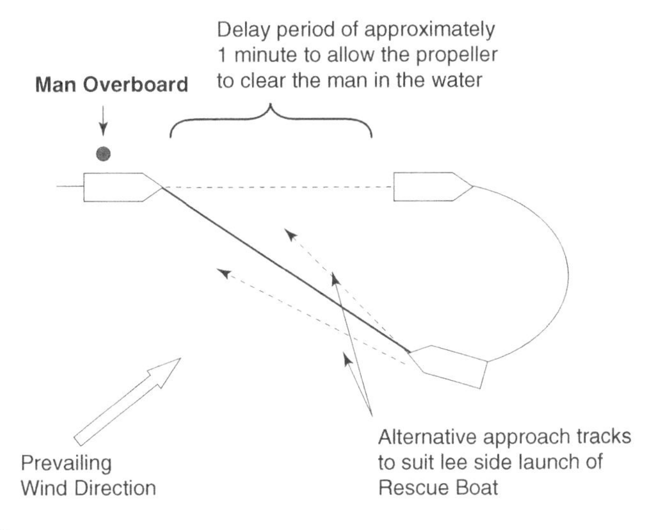

The Delayed Turn.

The single delayed turn is one which is exercised to delay helm action affecting the man in the water, assuming that the casualty is moving clear down the ship's side away from the propeller area. As the man overboard is swept past the propeller region the helm can be placed hard over to the opposite side to which the man fell. This action will cause the vessel to go into the first part of the ship's turning circle. As the circle is scribed, the ship's speed would expect to be reduced and engines would expect to have been placed on stand-by from the time of the alarm. The propeller(s) should pass well clear of the casualty, with this manoeuvre.

The vessel would line up the ship's head with the man in the water and make an approach suitable to create a lee to launch the rescue boat. The approach direction should take account of the prevailing wind direction to ensure that the parent vessel does not set down on the man in the water, while at the same time favouring the Rescue Boat launch.

The double elliptical turn

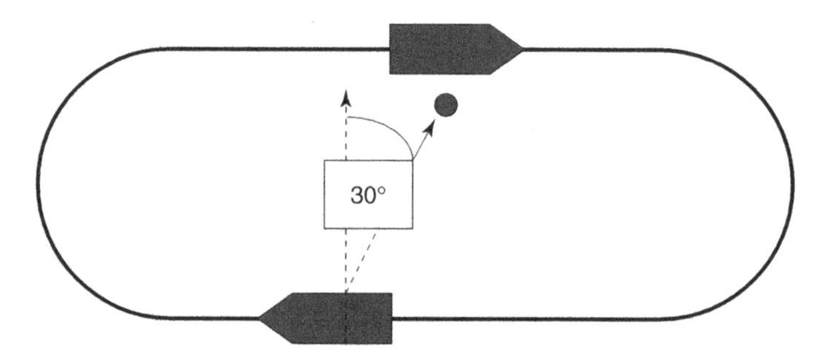

The double turn, as it is often referred, has a distinct advantage over the Williamson and Delayed type turns, in that the lookouts watching the man overboard do not have to change sides during the manoeuvre, but can to retain 'line of sight' on the man in the water.

Once the man is lost overboard the vessel is expected to turn towards the side on which the man fell and manoeuvre at reduced speed to a position to bring the casualty approximately 30° abaft the beam. Once this position is reached, the Rescue Boat can be launched on the vessel's lee side. Once the recovery boat is clear, the vessel can complete the double turn to recover both the boat and the casualty.

NB. *If the prevailing weather is such that recovery of the rescue craft is difficult, it may be necessary to generate a revised ship's heading to create a further lee to benefit the recovery operation.*

Man overboard – when not located

In the event that a Williamson Turn, or other tactical turn is completed and the man is not immediately located, the advice in the IAMSAR manual should be taken and a search pattern adopted. The recommendations from the manual suggest that where the position of the object is known with some accuracy and the area of intended search is small, then a 'Sector Search Pattern' should be adopted.

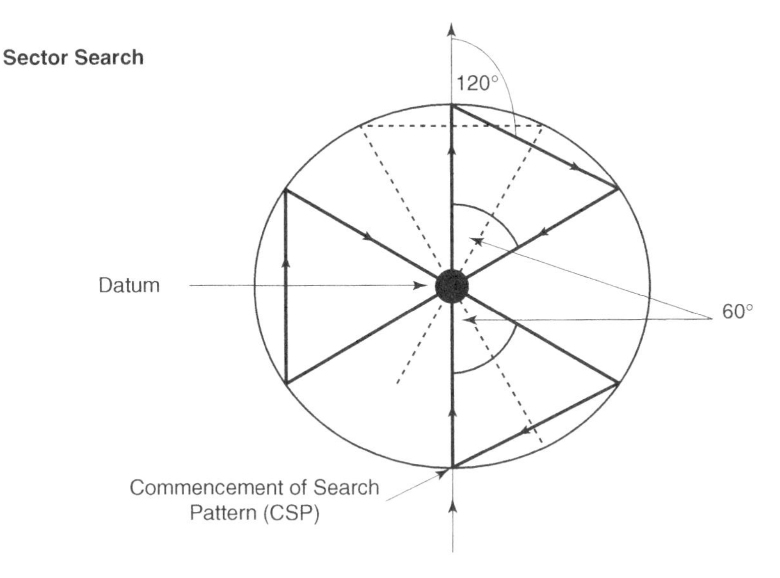

A Sector Search.

Suggested construction circle to commence the search pattern as the vessel crosses the Circumference

Track Space = Radius of Circle

Although a table of suggested track spaces is recommended in the IAMSAR manual, factors such as sea temperature, etc. can expect to be influential where a man overboard is concerned. In such cases, a track space of 10 minutes might seem more realistic with regard to developing a successful outcome.

NB*. Even at 10 minute track space intervals, at a search speed of 3 knots it would still take 90 minutes to complete a single sector search pattern.*

It will be seen that the alteration of course by the vessel is 120° on each occasion when completing this type of pattern. In the event of location still not being achieved after pattern completion, or in the event of two search units being involved, an intermediate track could be followed.

Search patterns – choice and aspects

Masters of ships called in to act as a search unit or who find themselves designated as an On Scene Co-ordinator (OSC) may find that SAR Mission Co-ordinator (SMC) would provide a search action plan. However, this is not guaranteed, and the choice of the type of search pattern to employ may fall to the individual Master.

Clearly, a choice of pattern will be influenced by many factors, not least the number of search units engaged and the size of the area to be searched. It will need to be pre-planned to ensure that all participants are aware of their respective duties during the ongoing operation. To this end the navigation officers of vessels can expect to play a key role within the Bridge Teams.

Establishing the search

Initially the Datum for the search area will need to be plotted. Where multiple search units are employed to search select areas, each area should be allocated geographic co-ordinates. This would reduce the possibility of overlap and time wasting, and assist reporting, by eliminating specific sea areas.

Once the search area(s) has been established and an appropriate pattern confirmed, the 'Track Space' for the unit or units so engaged must be established. This must be selected to provide adequate safe separation between searching units while at the same time taking into account the following factors:

a) The target size and definition.
b) The state of visibility on scene.
c) The sea state inside the designated search area.
d) The quality of the radar target likely to be presented.
e) Height of eye of lookouts.
f) Speed of vessel engaged in search operation.
g) Number of search units engaged.
h) Time remaining of available daylight.
i) Master's experience.
j) Recommendations from MRCC.
k) Height above sea level (for aircraft).

Additional influencing factors:

Night searches can be ongoing with effective searchlight coverage.

Length of search period may be restricted by the endurance of the vessels engaged.

Target may be able to make itself more prominent if it retains self help capability.

Pattern and respective track space should be selected with reference to the IAMSAR manuals and in particular Volume III.

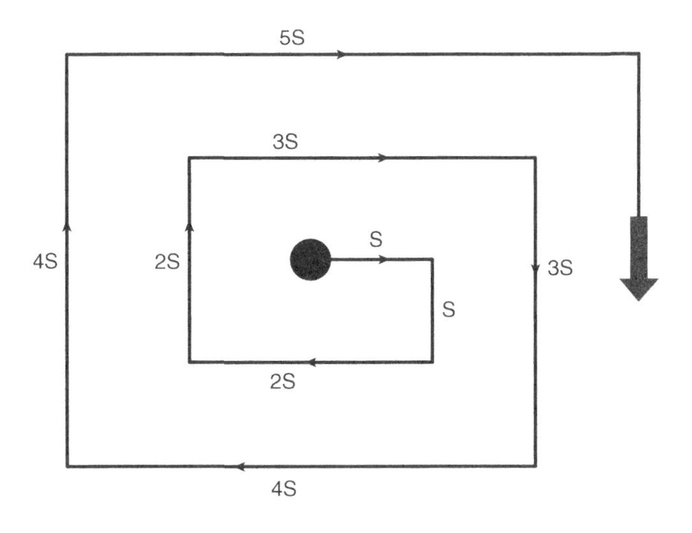

The expanding square search pattern.

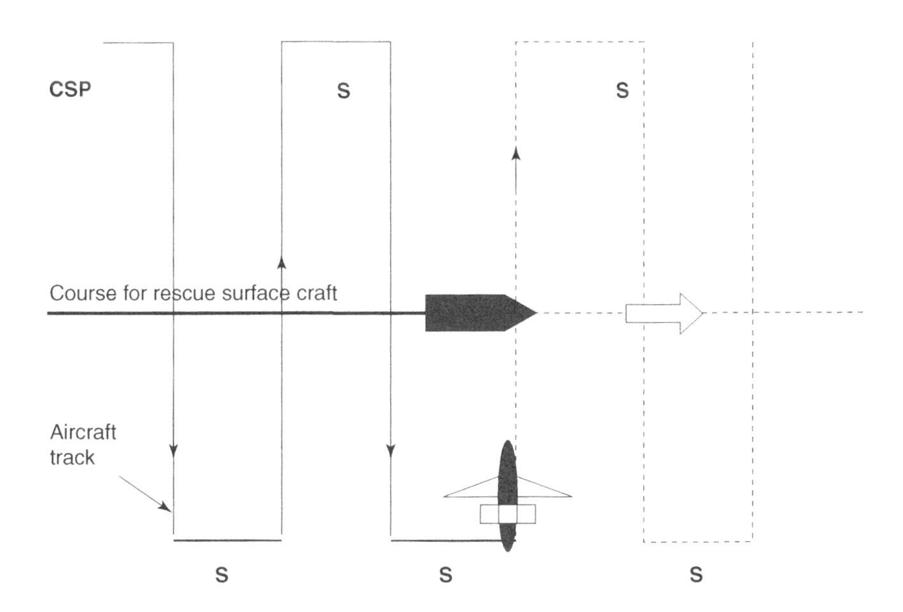

Co-ordinated surface search.

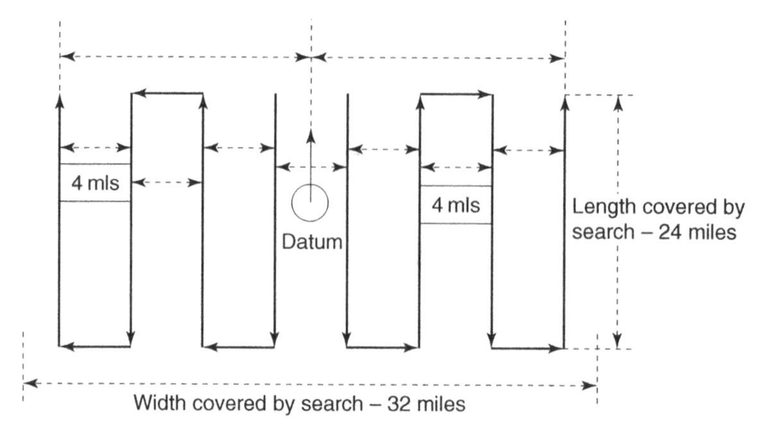

(a) Parallel search – Two (2) ships

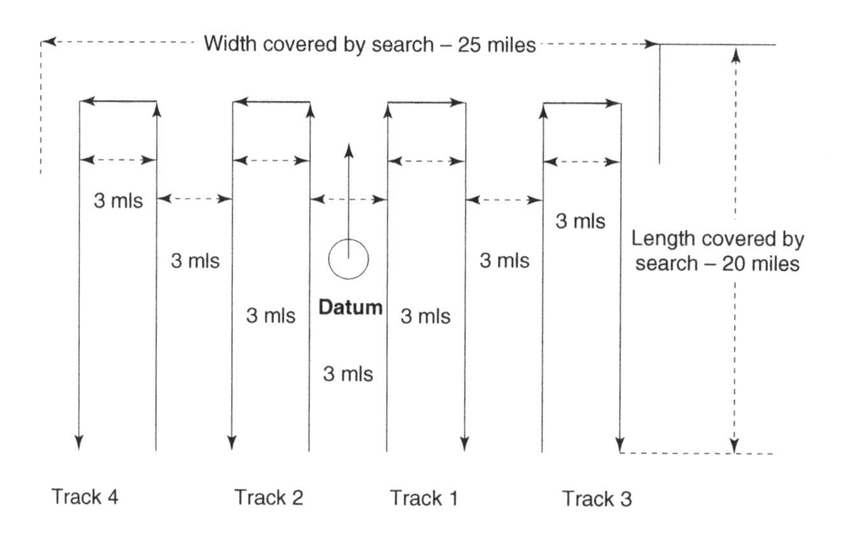

(b) Parallel search – Four (4) ships

Search pattern manoeuvres.
Arrow from Datum indicates the direction of drift.

The manoeuvre is conducted around a known 'Datum Position' with each perimeter being extended a further track space. The track space adopted needs to be practical, bearing in mind the circumstances and objective of the search; track space being reflective of the target definition, the sea temperature, state of visibility, height of eye, sea state, etc.

It should be borne in mind that it would be expected that conducting any search pattern would mean that a bridge team would be active. Also, other vessels may be in close proximity and the danger of collision must be a real consideration.

Working with helicopters

The combined operations of surface craft and aircraft has become much more common for both routine and emergency operations. More new and specialist tonnage is now being constructed with heli-deck landing facilities and can expect to engage with a variety of rotary winged aircraft. Ship's crews need to be trained to cater for land-on or hoist operations while ship's handlers need to appreciate the needs of the pilot and his/her aircraft.

Early communications with the aircraft would expect to confirm the rendezvous position and the local weather conditions; the ship's head being usually set about 30° off the direction of the wind. Hoist operations will invariably take place off the ship's Port side to accommodate the starboard access and the winch position of the aircraft.

A Puma helicopter engages in a pilot transfer to the Port deck side of a large oil tanker. Calm weather conditions prevail at force 3, and the sea area is clear of other traffic.

Deck preparation to engage with helicopters

Masters would be expected to put their bridge into an 'alert status' for any helicopter engagement and this would mean that the Master would 'take the con' of the ship, have engines on stand-by manoeuvring, and be operational with a full bridge team in place.

Depending on the depth of water, the use of the ship's anchors may or may not be appropriate, but should be considered once the rendezvous position and the subsequent approach plan has been established.

The ICS *Guide to ship/helicopter operations* provides this, but a brief resumé is included here for purpose of familiarity of the reader.

1. All rigging stretched aloft, all stays, halyards and aerials, etc. should be secured, lowered or removed to prevent interference with the aircraft.
2. All loose objects adjacent to and inside the operational area should be removed or secured against the downdraft from the helicopters rotors.

3. A rescue party with at least two men wearing fire protective suits should be detailed to stand by in a state of readiness.
4. The ship's fire pumps should be operational and a good pressure observed on the branch lines.
5. Foam extinguishing facilities should be standing by close to the operating area. Foam nozzles/monitors should be pointing away from the approaching helicopter.
6. The ship's rescue boat should be turned out and in a state of readiness for immediate launch.
7. All deck crew involved in and around the area should wear high visibility vests. Hard hats should not be worn unless secured by substantial chin restraining straps.
8. Emergency equipment should be readily available at or near the operational area. The minimum equipment should include:
   a) A large axe
   b) Portable fire extinguishers
   c) A crow bar
   d) A set of wire cutters
   e) First Aid equipment
   f) Red emergency signalling torch
   g) Marshalling batons at night.

   Most modern day vessels would also have up-to-date power tools available in addition to the basic emergency equipment.
9. Correct navigation signals, for restricted in ability to manoeuvre displayed.
10. Communications tested and identified radio channels/frequencies guarded.
11. The hook-handler (if applicable) is adequately equipped with rubber gloves and rubber soled shoes to avoid the dangers from static electricity.
12. All non-essential personnel clear of the area and the OOW informed that all preparations are complete and that the vessel is ready to receive the helicopter.

The 'Seaway Falcon' originally built as a drill ship, but later converted into a cable ship fitted with a dynamic positioning system.

Manoeuvring – following collision

In situations where a collision has taken place between two vessels, the subsequent action of each ship will be dependent on the circumstances. The types of vessels involved and the position and angle of impact will dictate who does what, and when.

Examples of this can be easily identified, particularly in the case of a tanker. For the other vessel to pull away prior to a blanket of foam being established over the contact area, this could well generate a high fire risk from tearing metal hulls apart. Another prime example can be highlighted where one vessel is embedded into another and provides an increased permeability factor. For the ship to withdraw, this would effectively remove the plug to the impact area and allow a major flooding issue to affect the impacted vessel. It may even be prudent for the striking vessel to retain a few engine revolutions, to ensure that the ships do not separate of their own accord and too soon.

Masters of vessels in collision are obliged, by law, to remain on scene and render assistance to each other. Therefore, the thought of turning away, without a legal exchange of information, would be deemed an illegal action. Circumstances, however, may make a sinking vessel seek out a shoal area to deliberately beach the ship, to avoid the total constructive loss, assuming the geography allows the beaching option.

It would, in probably every case of serious collision, be a matter of course to issue either an 'Urgency' or a 'Mayday' communication. Depending on response, each ship would probably need to be dry-docked or towed to an initial Port of Refuge. Again, the circumstances – such as the availability of engines, etc. – will influence subsequent actions.

Instances of collision require damage assessments to be made aboard respective ships. Provided the Collision bulkhead has held and tank tops are not broached the ship's stability could well be intact. If damage has occurred above the waterline this might be patchable. Where damage is on the waterline, the action of listing the vessel to the opposite side could bring the damaged area above the surface and prevent flooding. In the case of flooding from damage below the waterline, ordering the pumps onto the effected area may only buy valuable time, depending on the extent of the damaged area. Every case, every situation will have a different set of circumstances.

It is important to note that damage control on large ships is extremely limited. In most cases, manpower is short and resources are inadequate by size, if available at all. The incident will undoubtedly require the seamanship skills of the Master to either return the vessel to a safe haven or abandon the ship and order personnel into the second line of defence, survival craft.

Many collisions have occurred in poor visibility in both day and night time conditions. The status of vessels could change quickly from that of a Power Driven Vessel to being one which is disabled and needs to go to a Not Under Command Condition. As the reality of the situation comes to light, the weather conditions will have played a significant part and will continue to influence future outcomes.

Beaching

Beaching is defined as deliberately taking the ground. It is usually only considered if the vessel is facing catastrophe, which could result in a total constructive loss of the ship. A Master would run into shallows and deliberately take the beach with a view to instigating repairs to the ship, and with the intention of causing her to re-float and attain a Port of Refuge, at a later time.

The action of beaching a ship is considered extreme but the loss of the vessel would be considered far more dire. Once beached in shallows, the ship will not sink provided that the vessel can be held in position on the beach. Not an easy task, retaining position in such circumstances, especially on a rising tide condition.

Beaching should not be mistaken for 'running aground'. Beaching is a deliberate act compared to grounding which takes place by accident. When a vessel is beached, it is meant as a controlled activity – an activity where the type of beach is selected. This is unlike a vessel grounding which makes contact with the ground where elements and circumstances dictate.

Seemingly, a ship making contact with the ground – as in beaching or as in accidental grounding – results in the same predicament for the vessel. However, a controlled operation could distinctly favour less damage to the ship's hull. The selection of a rock-free, sandy beach could be beneficial when compared to grounding on a rocky surface.

Incident Report

One of the most recent incidents of deliberately beaching a vessel occurred with the container vessel 'Napoli' off the South Coast of England in January 2007. Following noted damage and loss of watertight integrity to the vessel, the decision to beach the ship in the Lyme Bay area was ordered. This operation, assisted by tugs, although generating the loss of several containers, allowed the majority of cargo and oil fuel to be salved, in what was a successful but lengthy period of salvage.

Grounding

A highly undesirable situation for any vessel to be in. Grounding is generally caused by poor navigation, possibly involving human error, or by machinery malfunction coupled with bad weather. In either case, the accidental contact with unselected ground could have serious consequences for the ship's well being.

The total loss of underkeel clearance for the vessel tends to occur with resulting contact with whatever surface is under the ship at the time. The benefits of ships built with double bottom (DB) structures can be clearly seen as a positive asset; bearing in mind that, if the outer ship's shell plate is broached, then the tank tops of the DB construction could prevent the flooding of the vessel. A ship can also float on her tank tops, provided these are not damaged.

Once a vessel has taken the ground, Masters should order a damage assessment to be made. Initially to check the watertight integrity of the hull; whether the engine room is in a wet or dry condition; if the incident has generated any casualties; or if the ship is causing any pollution, etc. Subsequently, a full set of internal tank soundings should be taken to ascertain the state of the internal structure. Also a full set of external soundings around the ship's hull should be made to gain positive information regarding the ground that the vessel has made contact with.

Beaching diagram

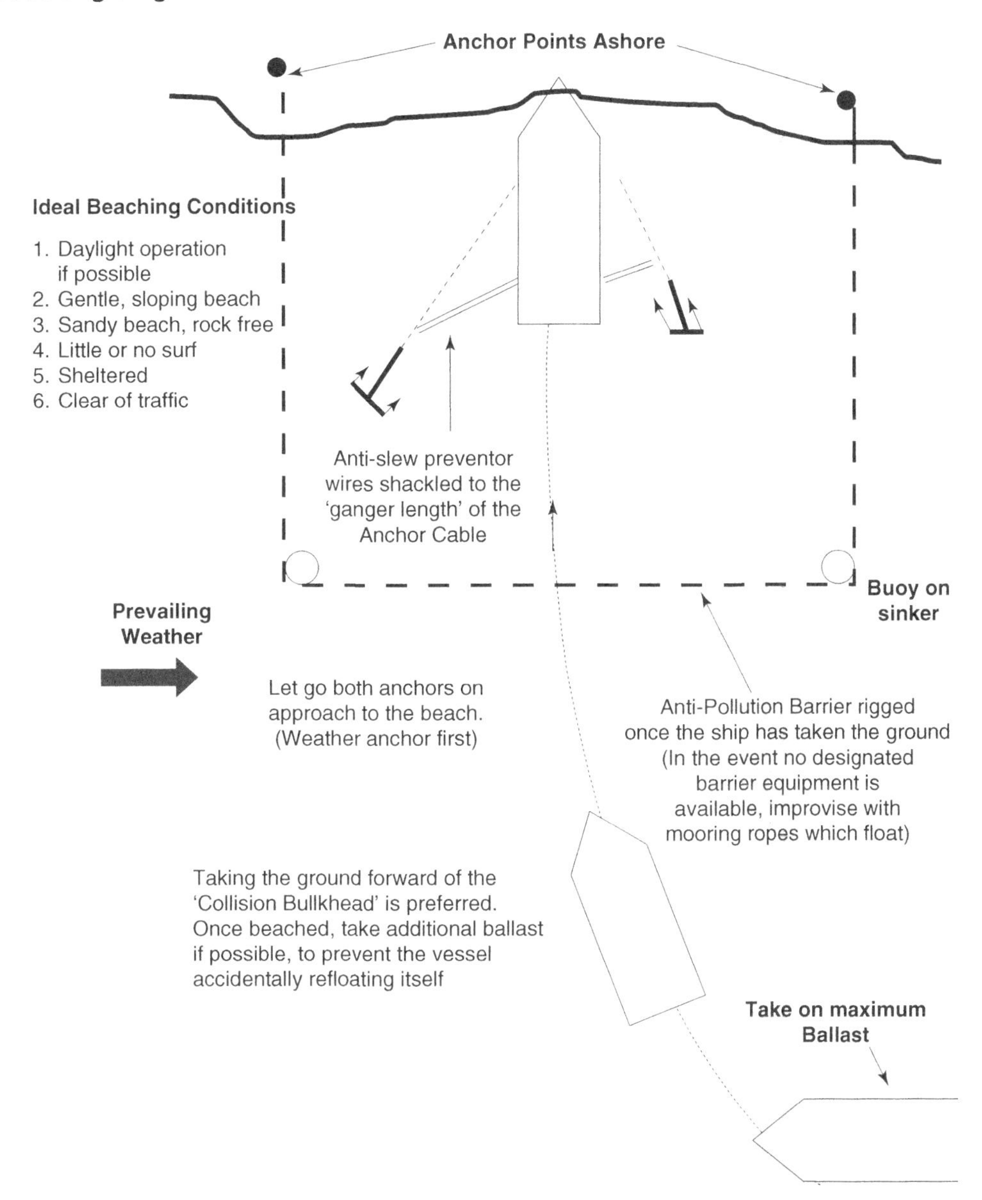

Actions following beaching or grounding

In both the incidents of beaching and grounding, the ship's anchors should be walked back with the idea to prevent the vessel accidentally re-floating itself on a rising tide before shipboard personnel are ready to attempt a controlled re-float operation.

When initial repairs have been instigated and completed to ensure that the vessel will not sink, preparations to re-float the vessel can be made. The tidal data should be consulted to consider a suitable day and time to re-float the ship. Prior to re-floating, arrangements must be made to bring in a stand-by vessel as a second line of defence in case additional damage is caused when moving the vessel astern into deep water.

Ballast, initially put in to make the vessel as deep as possible, should be pumped out to lighten the vessel and adjust the ship's trim to suit the angle of the ground.

A flat bottomed dredger is blown ashore after losing her main engine power, onto a sloping sanding beach off a North West English coastline. Stay wires are seen, rigged to hold the vessel in position, to ensure she is not carried further inland on spring tides.

The vessel off the Lee Shore

Every shipmaster's nightmare is being caught off the lee shore with either no engine power or disabled steering gear. In either circumstance, the end product could be that the vessel becomes stranded and takes the ground, often leading to a total constructive loss.

With the loss of any of the essential navigation equipment it must be anticipated that the Watch Officer would call the Master who would be expected to take the 'conn' of the vessel. Clearly the circumstances of each case will be influenced by the prevailing weather conditions, especially the strength of the wind. The feeling of helplessness will inevitably prevail once a Chief Engineer informs the Master that the engines are beyond the repair stage or that the Rudder has broken away.

Different situations would call for respective actions by the Master even if these fall back to mere delaying tactics. Emergency procedures must inevitably be put into place and these must include any or all of the following:

1. Display the Not Under Command (NUC) signals.
2. Obtain an immediate weather report.
3. Place a position on the chart.
4. Instigate an urgency and/or a distress signal depending on circumstances.
5. Prepare anchors for immediate use and have an anchor party standing by.
6. Stand-by anchors for deep water anchoring.

7. Monitor the rate of drift effecting the vessels movement.
8. Turn out survival craft ready for a second line of defence in the event that the parent vessel becomes no longer sustainable.
9. As soon as the situation develops, the possibility of instigating repairs must be actioned as soon as practical after occurrence (one would expect this to have been carried out at the earliest stage following an initial damage assessment and it is assumed that repairs are not possible which results in the necessity for such emergency actions to take place at this time).
10. Communications should include contact to tugs, with the view to establishing a contract of tow.
11. Prepare a towline securing arrangement where tug operations are anticipated.
12. Depending on the stage of advancement of a stranding situation and the weather conditions a request for helicopter evacuation may have to be considered.

Delaying tactics, as previously mentioned, may prove a positive action when awaiting the arrival of tug(s) or helicopter assistance. In any event, walking back anchors could generate a hang up situation to hold the ship off the shoreline. Deep water anchoring may cause loss of the anchor and cable but this would be a small price to pay for bringing the ship home to an eventual safe haven.

Where steering gear has failed, through loss of rudder, use of twin screw action (assuming engines still available) could assist to cause emergency steering. However, not all ships have the benefit of twin propellers and an improvised jury steering (possibly by use of anchors) may suffice temporarily. A further option to use engines to 'stern bore' into the wind may also prove a possible delay tactic to allow time for tug assistance to become established.

Masters may be forced by circumstances to make difficult decisions including the one to abandon the vessel. If such action becomes necessary the decision should be made sooner rather than later as this provides more time availability to carry out a safer operation. Where possible, Masters should not abandon the vessel without leaving towing springs secured in positions fore and aft. Weather conditions may not permit such an activity at both ends of the vessel, but even at one end, it may provide means for a tug unit to hold off until the weather abates.

Sea anchor use

The principal use of a 'sea anchor' is to hold the bow into the direction of the weather and this may be usefully employed in the event of the ship losing power off the 'Lee Shore'. Survival craft tend to have small, purpose-built sea anchors, to act as drogues or control elements for the handling of small craft in bad weather. Whereas the larger size vessel would need some form of improvisation to be effective in holding the ship's bow into the wind.

Few vessels, if any, are fitted with a purpose-built construction to act as a drogue, but most ships could generate some form of improvised drag weight to reduce the rate of drift on the vessel. Examples may be found in mooring ropes trailed ahead of the vessel or even left in a coil, then streamed and secured from either bow.

This is one of those cases where the means justifies the end. Any object, floating or partially submerged, trailed from the bow region, can be expected to reduce the drift rate. At the very least it may buy additional time to allow tug(s) to arrive on the scene.

The sea anchor is a 'jury rig', an improvisation at best. They are invariably difficult to rig and deploy in exposed conditions. The effectiveness is questionable on the type of sea anchor that can be deployed. Walking back the anchor(s) below the keel may work just as well, depending on weather and sea conditions and the forward construction of the vessel. The overall performance of a sea anchor could possibly be enhanced by prudent ballasting at the fore end. This would probably cause the bow to lie deeper with increased draught, while the stern superstructure becomes more exposed. Overall this may allow the vessel to weather-vane and allow the sea anchor to become more effective.

Stern bore into wind

An alternative action to the use of anchors, when off a 'Lee Shore' may be available if the vessel still has use of engines, as with a situation where the rudder has been lost. A typical 'stern bore' manoeuvre, for a right hand fixed propeller vessel is illustrated below. When the engines are going astern the transverse thrust effect would normally take the stern to port. However, the pivot point moves aft and the longer forward part of the vessel will act as a flag and line up downwind. The pivot point acting like the position of a flag pole.

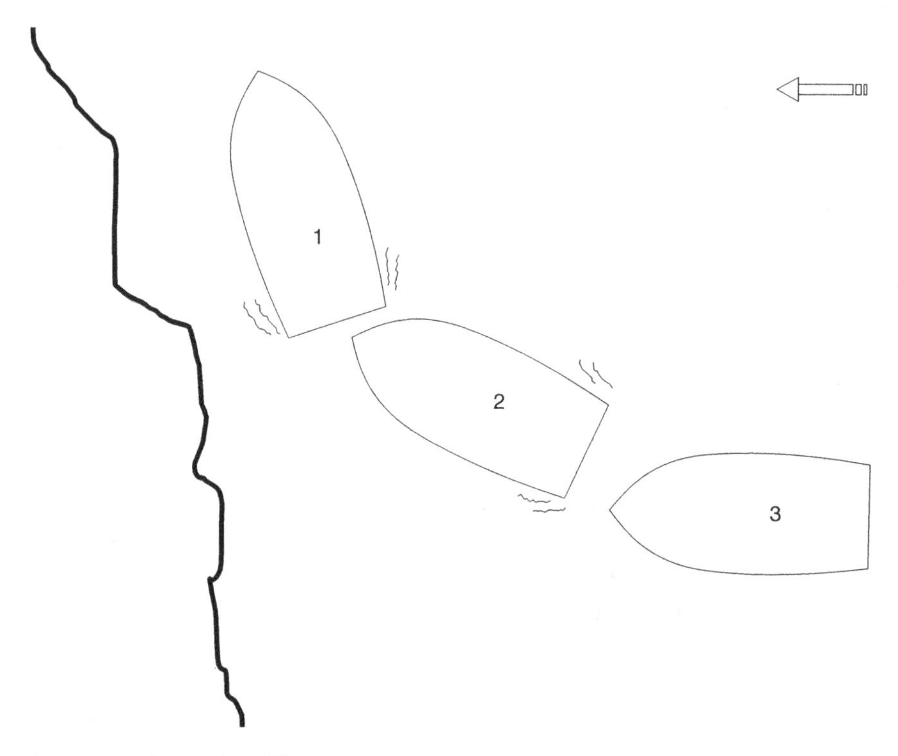

Stern bore away from a Lee Shore.

The scenario of stern bore into the wind can be achieved with ships with accommodation all aft or all forward, but would probably be difficult, if not impossible, with high sided vessels like car carriers and passenger vessels.

It is imperative if conducting this manoeuvre that adequate sea-room is retained and kept available, to take account of any leeway influences.

Manoeuvring in ice conditions

Working vessels in ice conditions is never an easy task and can become extremely dangerous. The prevailing ice condition, namely the coverage, the thickness and the type of ice prevalent will undoubtedly influence the behaviour and handling of the vessel. Once inside the ice limits, Masters could expect to double watches and proceed with caution at a reduced speed. Continuous monitoring of the weather reports and specific ice reports is considered an essential element to the prosecution of the voyage.

Depending on location, i.e. Canadian waters, Baltic waters or the Arctic or Antarctic regions, will depend on the sources of reference for ice reports. One of the main sources for the Northern regions is the International Ice Patrol (IIP). In the case of the Baltic, Keil Harbour Radio provides updated information on request. Southern Hemisphere, ice information is available from Punta Arenas and relevant observation stations around Antarctica.

NB. *Additional references can be made from respective Admiralty Sailing Directions regarding communication links and relevant ice information.*

Progress in ice conditions is, by the very nature of the environment, slow. On occasions where heavy ice prevails, the progress may even be astern or in another direction other than ahead. Ships hoping to proceed should have 'Ice Classification' with a high grade of Ice Strengthening. Any movement in ice will generate ice damage to the vessel and Masters and crews must be prepared for such damage. Particular areas of the vessel which are susceptible to damage include:

The waterline region
- Fresh water and ballast tanks freezing and cracking;
- Upper rigging parting with the added weight of ice accretion;
- Seawater intakes and discharge outlets freezing up causing overheating of machinery;
- Rudder and propeller areas suffer impact damage from large ice formations;
- Deck machinery seizes up due to extreme temperatures;
- Pipe lagging distorts and cracks allowing pipes and joints to freeze and crack.

Protective measures can be taken to prevent some of the above, like adding salt content to Ballast Water to reduce the risk of freezing, maintain deck machinery in a continuous 'run mode', clear seawater filters each watch, steam clear ice accretion, etc.

However, at the end of the day, anchors may still stick in the hawse pipes when they are required and the extreme cold climate is just not compatible with smooth operations.

Watchkeeping practices – approaching ice regions

The navigation of the vessel inside ice waters requires maximum input by all those involved in the watchkeeping arrangement of the vessel. This is not limited to the Navigation Officers but must include the Engineering Personnel and the full complement of the crew. The 'Lookout' and the 'Quartermaster' can expect to play vital roles within the bridge team formation.

It must be emphasized from the onset that the prime function of watchkeepers inside the ice regions is that of keeping an effective lookout both visually and by radar. The ship's Master may consider it necessary to double watches once the vessel is known to be inside ice limits during the ice season. The main purpose of a second bridge watchkeeper would be to provide the vessel with a continuous 'Radar Watch' as and when deemed necessary. Neither should it be considered unusual to stop the vessel's motion during the hours of darkness when inside ice limits. This should be especially recognized when it is realized that all ice targets make 'poor' radar targets.

The doubling of watches, and the strategic positioning of 'lookout personnel' should not be taken lightly. Double watches is never popular, nor leaving lookouts exposed for lengthy periods. However, the main concern is for the safety of the vessel, and the needs of the single individual take second place. Such practice should reflect the adverse conditions and Masters are advised of the dangers incurred by 'fatigue' affecting the ship's watchkeepers.

Clearly, evidence of ice detection would be prominent from the telltale sea-temperature dropping below the zero level. Together with geographic location coupled with Navtex, radio communications and satellite imagery, a Master would expect to be reasonably forewarned of when and where his vessel has entered ice infested waters. To this end, the prudent Master would have briefed his Watch Officers and would be proceeding with utmost caution to pass through the threatening area.

Most companies would expect to have 'Company Standing Orders' in place, acknowledged by all officers on joining respective ships. Each individual vessel would also proceed under the Night 'Standing Orders' of the ship's Master. These orders would expect to reflect the legal obligations stipulated by SOLAS, that: **All ships on being notified that dangerous ice is on or near the intended track, should alter their course and proceed at a moderate speed at night. It should also be noted that any ice sightings, for which no report has been given, should be reported to the Authority, via the nearest coast radio station.**

Example: Master's night standing orders (approaching or inside ice limits)

1. The Officer Of the Watch (OOW) is expected to call the Master, as per Company Standing Orders, on the sighting of any dangerous ice, or in the event of any emergency situation that is deemed necessary.
2. Inside known 'Ice-Limits' the ship's engines will be in stand-by mode and the ship must proceed at a moderate speed, taking account of the prevailing conditions.
3. A continuous lookout is to be maintained throughout the watch period by both the primary and secondary lookouts; the lookout being maintained by all available means inclusive of visual and radar methods. The OOW will consider himself as the prime lookout throughout the watch period.
4. The OOW will have full control of the navigation and manoeuvring of the vessel in the absence of the Master and should not hesitate to alter the vessel's course or speed to pass clear of any apparent danger.
5. The ship's position should be monitored at regular intervals and in any event should not exceed fifteen (15) minute intervals in coastal waters or inside known

ice limits. This position should be corroborated by use of the echo-sounder whenever possible.

6. Weather conditions should be monitored throughout the watch period and in the event of any adverse change which could affect the vessel's performance the Master should be informed immediately.
7. The vessel will be on Manual Steering while inside known ice limits.
8. A continuous radar watch is to be maintained at peak performance throughout the watch period. All radar targets are to be systematically plotted and should a close quarters situation be developing, the Master should be informed immediately at the onset.
9. In the event of restricted visibility being experienced, the Master should be informed immediately and the Prevention of Collision Regulations adhered to.
10. On sighting any ice, the position and full description of such ice should be noted and the Master informed. A full account of all ice sightings should be noted on the navigational chart and recorded in the Bridge Log Book.
11. The OOW should maintain a continuous listening VHF radio watch throughout the period of duty and take specific note of all ice and associated weather reports, informing the Master of any and all adverse elements.
12. The OOW should consider himself as the Master's representative while holding the duty watch and should not hesitate to call the Master in the event of any hazard or concern that may stand the vessel into danger.

Berthing in ice

The Baltic Eider moored nearly alongside, starboard side to, stern to the berth, in Helsinki. Where possible, tugs or ice breakers tend to clear pack ice from berthing areas in order to allow vessels to draw fully alongside. Where ancillary craft are not available to clear the ice, the vessel itself can sometimes manoeuvre to use propellers and bow thrust slipstreams to clear floating ice, but such action is not always successful.

The dangers from ice can affect the safety of the ship even when alongside, especially where the vessel is on a river berth with current or tidal action. The natural flare of the bow permits any moving ice in the river to flow between the bow and the quay. This can be exacerbated if ice breakers move up river to break up ice accumulations. The floes move downstream in the current and add further weight to the 'ice wedge' between the quayside and the hull. Such action can force the vessel to part its moorings and be torn from the berth.

Experience has shown that mooring the vessel by means of the anchor cable – or Insurance Wire – provides no guarantee that the vessel can be retained alongside against an ever increasing accumulation of ice between the ship's hull and the quayside. Normal mooring ropes can stretch and may have some endurance in such conditions; however, anchor cables and insurance wires have little elasticity and may be the first to break under the weight of ice and current, in the build up.

Specialist ice operations

The 'Bransfield' seen moored alongside an 'Ice Shelf' in Antarctica. Gangways are seen deployed to the ice surface while moorings are stretched to inland anchor points. Ice cap research stations rely on such vessels for bulk supplies and relief operations. Emergency relief activities being generally handled by helicopter transport.

These vessels are of Ice Classification and are generally fitted with an ice breaker bow design. They can expect to encounter close pack ice and open pack ice as well as virtually every type of ice formation during their tour of duty. Coming alongside such an ice feature as an ice wall of an ice shelf has inherent dangers of residual ice floes being blown into obstructive positions. Also, the shelf may have under surface projections. A close inspection of the intended berthing position and slow approach is standard practice for ship handling personnel. Sharp and jagged edges of an ice wall are to be avoided if drawing alongside.

Ice breaker towing activities

In ice infested regions, especially where the winter season is severe, the ice thickness can cause many problems for commercial traffic. The lack of power, or an ice breaker bow, incorporated into many designs of merchant shipping examples leads to such vessels becoming 'nipped' in ice formations and their progress brought to a halt.

Most ice regions – like the Gulf of St Lawrence and the Baltic Sea – operate national ice services to maintain traffic movement around respective coastlines and harbour entrances. Clearly, effective communications are essential to bring in ice breaker assistance and these vessels, once involved, have the ability to break channels in towards and around the effected vessel. They can also provide towing facilities for low powered vessels or disabled vessels often being equipped with both bow and stern towing facilities.

A ice breaker engaged in towing a small coastal vessel in the Baltic Sea in 2003. The movement is slow and close up, to allow the towed vessel to gain the benefit of the clear water being exposed by the lead of the ice breaker. The towing operation is ongoing through light broken pack ice on approach to Helsinki, Finland.

Ice convoy operations

Where access to ports and harbours is restricted due to ice accumulation, it is not unusual to see ice breakers commanding merchant ships in a convoy formation. The Officer in Charge of the ice breaker leading the convoy could expect to contact all vessels within the convoy and ascertain relevant details of each vessel, inclusive of maximum speed, draughts and communication channels. The speed of the convoy will be dictated by the prevailing ice conditions, state of visibility and the speed of the slowest vessel in the convoy.

It is anticipated that the lead vessel – the ice breaker – will generate an appropriate heading into the ice field for the convoy vessels to follow. However, continuous movement is not always guaranteed and ships may become 'beset' in the ice. This may require the ice breaker to double back to break the ship free, or tow her out into clearer waters.

NB. *Vessels participating within an 'Ice Convoy' would generally be required to have suitable towing arrangements readily available in the eventuality of being trapped in ice formations.*

The sequence order of vessels in convoy will be dictated by the commander of the ice breaker, but will tend to vary with the general particulars of each vessel; the motive power and the beam width being significant in influencing position. The distance apart between respective ships is recommended to be a suitable distance to gain the benefit of the broken ice ahead without generating an unnecessary collision risk. Such a distance will be found by taking account of the overall speed of the convoy and the experience gleaned from the actual movement of the vessels in the prevailing ice conditions. This distance is unlikely to be less than 150–250 metres apart, allowing enough distance for the following vessel to operate astern propulsion without causing a close-quarters collision situation.

Vessels in convoy should have effective communication facilities to reflect accurate movement, inclusive of sudden orders to stop and/or moving astern. Such communications would include VHF and the use of the International Code of Signals as well as sound signalling apparatus. The use of search lights by vessels in ice conditions must also be anticipated.

Vessel movements would be under manual steering control, lookouts would be posted, and main engines would be operating on manoeuvring speed. Radar observation would be on short range. Tight position monitoring, by primary and secondary position fixing systems, should be employed wherever possible.

Navigation – ice accretion risk

The possibility of a vessel being affected by ice accretion is always present in cold climates in the high latitudes. Certain ships are more prone than others to these affects, i.e. container vessels, fishing boats, high-sided car carriers and ships with deck cargoes.

The increased weight, high up on a container stack or in upper rigging, can and will add considerable weight to the vessel in a position above the vessel's Centre of Gravity 'G'. Such added weight will effectively cause G to rise towards the Metacentre 'M', so reducing the ship's GM. The risk of G rising above M, causing a possible unstable condition for the ship, becomes a real threat to the positive stability of the vessel.

The Master must consider re-routing the vessel, if possible, to warmer latitudes. He should also reduce speed while in a cold environment to lessen the possibility of the chill factor generating ice accretion. Clearly, it is not always practical to alter course away from cold climates and it may become necessary to order crew members to remove ice formations from the ship. This is a highly dangerous task and must be exercised with extreme caution. A reduction in speed will also effectively reduce the effects of sea spray, which is one of the main sources causing ice accretion.

The alternative might be to compensate for the added weight of ice by adding ballast to lower tanks. This, of course, is only possible where lower tanks are available for filling and the action, depending on circumstances, could also generate detrimental free surface effects.

It is essential that the upper weight of ice is removed or compensated for to keep the vessel stable. Removal by crew is a hazardous task and personnel must be briefed as to the hazards and then properly equipped for the task; protective clothing, gloves, boots and hard hats being the order of the day. A risk assessment would need to be carried out prior to the task being undertaken.

Use of axes and shovels are considered standard tools for the task but steam hoses can be more effective if available and correctly employed. Ice overhangs above deck structures are particularly dangerous and should be removed in small sections where possible. Broken ice at deck level must be removed overside when safe to do so.

Ice accretion is associated with sub-freezing air temperatures and windy conditions and if experienced by a vessel the Master would be expected to report his position and the conditions.

Emergency (quadrant) steering

Previously, vessels fitted with 'quadrant steering gear' always had the availability of an auxiliary manual steering method and a mechanical emergency method of turning the rudder. The auxiliary was either controlled from a position on the poop deck in a three island ship design, or coupled to a large wheel inside the steering flat. In either case, each method could be integrated into the worm and pinion gear to physically turn the rudder. The large wheel inside the steering flat was set aft of the pinion cog and could be engaged in place of the worm gear; the worm gear being hinged to permit removal to allow the large manual wheel to be engaged.

This particular method of turning the rudder was laborious and would normally take at least two men to physically turn the large wheel, to generate rudder movement. A communications link was a permanent feature of the steering flat and orders from the conning position were passed to the helmsmen in order to comply with desired headings.

An alternative arrangement was mounted on the upper poop deck which would have connecting rods from the steering wheel position down to the worming gear. The linkage would be dog clutched to physically turn the worm arrangement, causing the quadrant to move to port or starboard. This steerage position was normally fitted with a small binnacle arrangement and magnetic compass, together with a communication link. Effecting transfer of the steering control from the navigation bridge to the remote aft position was achieved comparatively quickly by the ship's engineers. This was achieved by first removing the link pin from the telemotor control rod and inserting the connecting rods from the manual steering wheel above the poop; the link pin being common to both the telemotor system and the auxiliary system, in order to effect movement of the worm gear.

The remote steering position must be tested and logged in the official logbook at intervals of not more than 3 months.

Incident Report
In 1970, the three month emergency steering test was conducted aboard a twin screw vessel the Albany. The transfer procedure of control, from the navigation bridge to the aft steering position, was carried out well and the test was completed successfully.

However, after the ship's Master had ordered the test complete and control be returned to the bridge, the link pin sheared during removal from the dog clutch arrangement. With this common interconnecting pin broken, the engineers could not reconnect the steering system to the telemotor transmission.

The problem was resolved in that an engineer was ordered to make a replica pin in the engineering workshop, a task that took about 30 minutes. However, during this period of time the vessel was left without any form of steerage control. The Master immediately ordered the vessel to go to a 'not under command' status and display the NUC signals. He also ordered the engine room to stand by and reduced the ship to manoeuvring speed, adjusting the r.p.m. to steer the vessel by engines (being a twin screw vessel, this was possible). After the new connection pin had been manufactured, the steering gear was reconnected to the navigation bridge and the incident entered into the official log.

Auxiliary/emergency steering

The alternative position of steering the ship in the event that the main systems have been rendered inoperable varies in type. The most popular method would seem to be secondary hydraulic oil tanks and pumps, situated in the steering flat itself. These can be connected quickly to replace the bridge transmission system and can be operated by manual button controls, via the communication system. Older ships may still be fitted with mechanical means of moving the rudder via use of the aft mooring/docking winches (as illustrated below).

Quadrant steering

Quadrant steering has been virtually totally superseded by electro hydraulic systems but its basis of operation is worthy of note. It was also easy adapted for use as emergency steering from the steering flat.

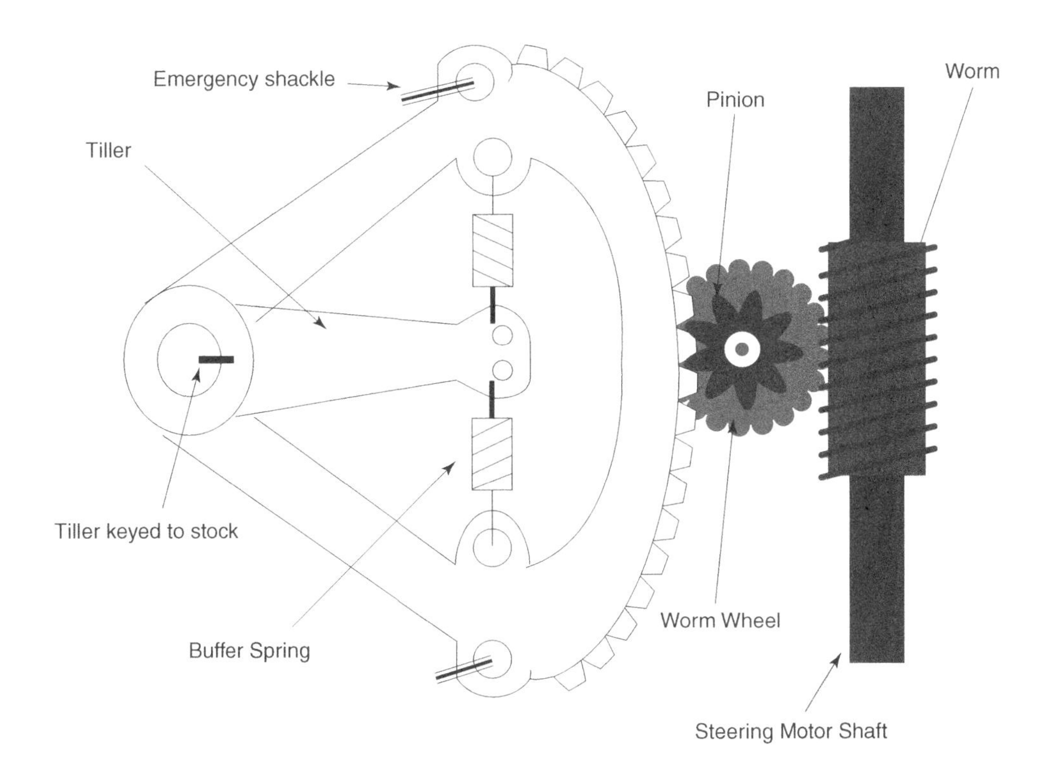

Emergency mechanical quadrant steering

Quadrant steering systems were usually fitted with emergency steering shackles secured to each end of the quadrant. These allowed heavy duty tackles (carried in the steering flat) to be connected in such a manner as to be able to heave the quadrant from side to side, so moving the stock and subsequently the rudder from port to starboard. The method employed the services of the stern docking winch, which operated on both quarters of the vessel.

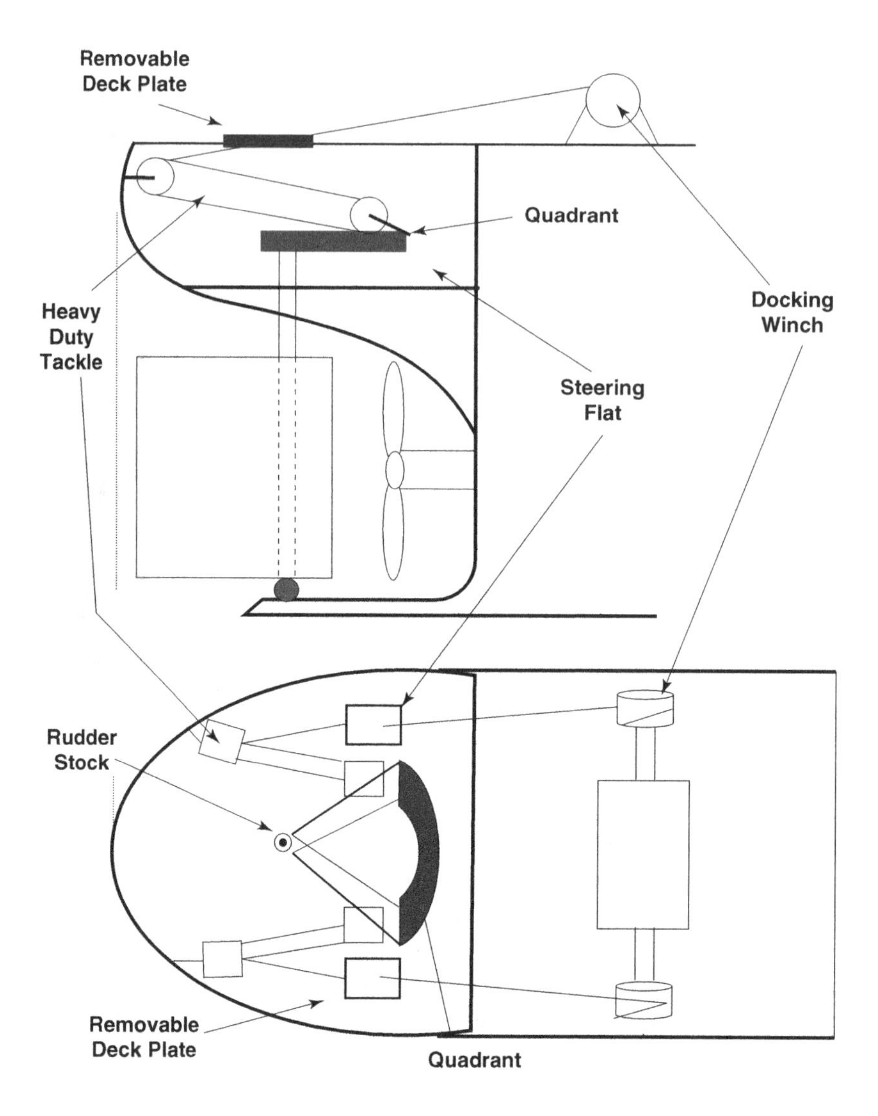

Jury steering

A jury rig is a term which suggests improvisation when a main component is lost or unavailable for use in the normal manner. Examples of this are seen in the case of a lost rudder, where 'Jury Steering' becomes an improvised method of steering the vessel.

Jury steering can be established in several ways, possibly by using drag weights from either bow or alternatively netting oil drums to stream from one quarter to the other acting as a drogue, to affect the ship's heading.

Steering by means of engines in twin screw vessels is not the normal use of a ship's propellers, and could be considered an improvisation where the designated steering gear is non-operational. Similarly, a small tug without the capability to pull a large vessel could be used aft, instead of forward to effect steering from an aft position.

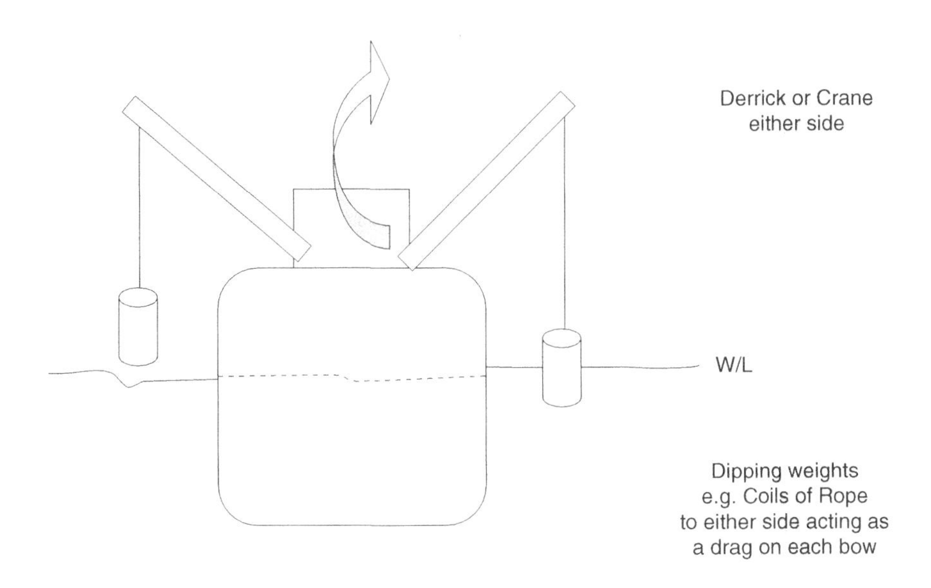

As an alternative to using coils of mooring ropes, anchors may be a suitable option, especially where the vessel is not operating with a bulbous bow arrangement. The dragging of a drogue of one form or another, like netted oil drums, over the stern quarters could also expect to influence the ship's heading. In the example above, the movement of the drogue from port to starboard would be caused by use of the port and starboard mooring winches.

Breakdown when engaged in gyro steering

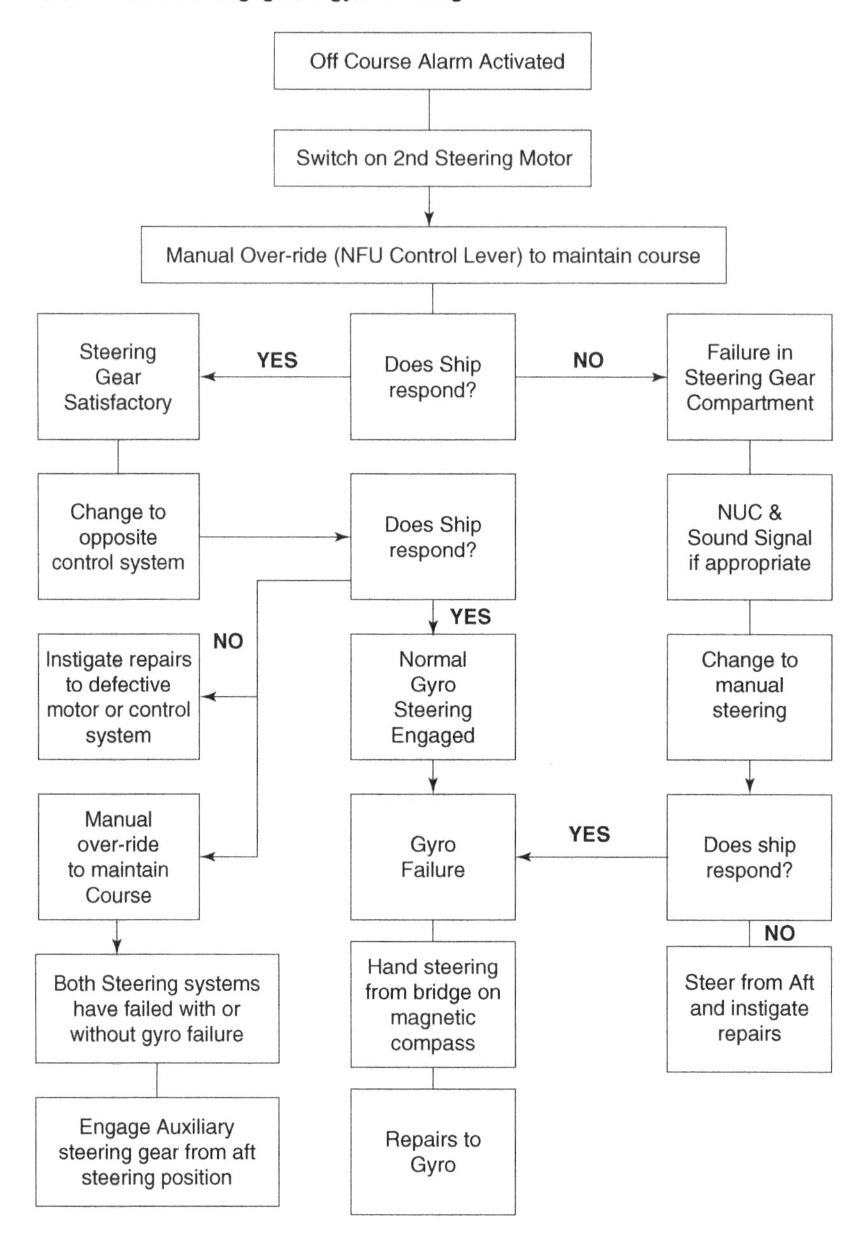

Steering gear failure – fault finder.
1. Additional manpower will be required to engage aft steering position.
2. Wheel should be placed at midships before changing over from bridge to aft.

Appendix A: Controlling elements of ship handling

Introduction; navigation bridge; steering gears; control systems.

Introduction

The many advances in ship manoeuvring aids are not difficult to see in this day and age. They are highly visible in all forms of the media. What is not readily seen is the means of controlling the new ideas within the practical environment and in positive applications. Development of the hardware has been matched only by the advances in control systems for an ever demanding and expanding maritime industry.

Solid state systems generated an unbelievable change across virtually all marine operations, not just in the topic of ship handling. However, ship handling from the integrated navigation bridge and high tech computer based control room, saw first hand benefits from product research and development.

Labour saving and user friendly systems have come to the fore and are delivering reliable, accurate sensing elements, of all shipboard parameters. Fibre optics, miniaturization and nano technology are continuing to change the way we operate within our marine environment. Fail safe systems have become the standard with monitoring of all elements becoming one of the most effective building blocks of any industry.

The navigation bridge

Virtually all modern ship design is now incorporating the integrated bridge. This has been a natural progression with increased technology in the fields of the Electronic Navigation Charts (ECDIS), Automatic Radar Plotting Aids (ARPA), Automatic Identification Systems (AIS), Voyage Data Recorders (VDRs), etc. Interfaces from speed monitors, echo sounders, Global Positioning Systems (GPS) helm and course recorders, etc. have generated the need for serviceable controlling spaces.

An example of an integrated bridge layout of a modern passenger vessel. The essential function of the lookout is not forgotten amongst the elements of technical innovation, with large exposed bridge windows ranging from port to starboard. Central controlling functions for steering and engines are enhanced by integrated ECDIS units, communications consol and all functional services set into an open user friendly environment.

Ship's steering gear

All cargo vessels over 500 grt are required to be fitted with power-operated steering gear and where the size of the 'Rudder Stock' exceeds a diameter of 250 mm (measured at the tiller) they must also be fitted with a power-operated auxiliary steering gear. In the case of a passenger vessel, they must also carry main power-operated steering gear and, where the rudder stock exceeds 230 mm, then a power-operated auxiliary steering gear must be fitted as well.

The operation of the respective steering gears must provide independent means of moving the rudder from 35° from one side, to 35° on the other, when the ship is at her maximum service speed. The main steering gear on all ships must be established on the navigation bridge and a remote alternative steering position away from the bridge site must be provided for the auxiliary system (usually in an aft position, e.g. the steering flat).

The movement of the rudder from 35° on one side to a position of 30° on the other side should not take longer than 28 seconds at the ship's maximum service speed. Power-operated steering systems are provided with limiting stops to prevent the rudder exceeding maximum angles of helm. Should a power failure occur, affecting the electric driven pumps of say an electro hydraulic steering system, provision for a manual operation or other alternative arrangement to turn the rudder, may be included.

Steering gear located at the remote station away from the navigation bridge must be tested at least every 3 months. A record of such tests being recorded by the Master in the ships 'Official Log Book'. The ship's main steering gear, located on the navigation bridge, must be tested prior to the ships departure from any port. Every such inspection and test would be entered into the 'Deck Log Book' together with a statement of fact that no defects were observed. Such tests would include indication that the helm indicator and the rudder indicator are seen to function correctly.

Telemotor transmission

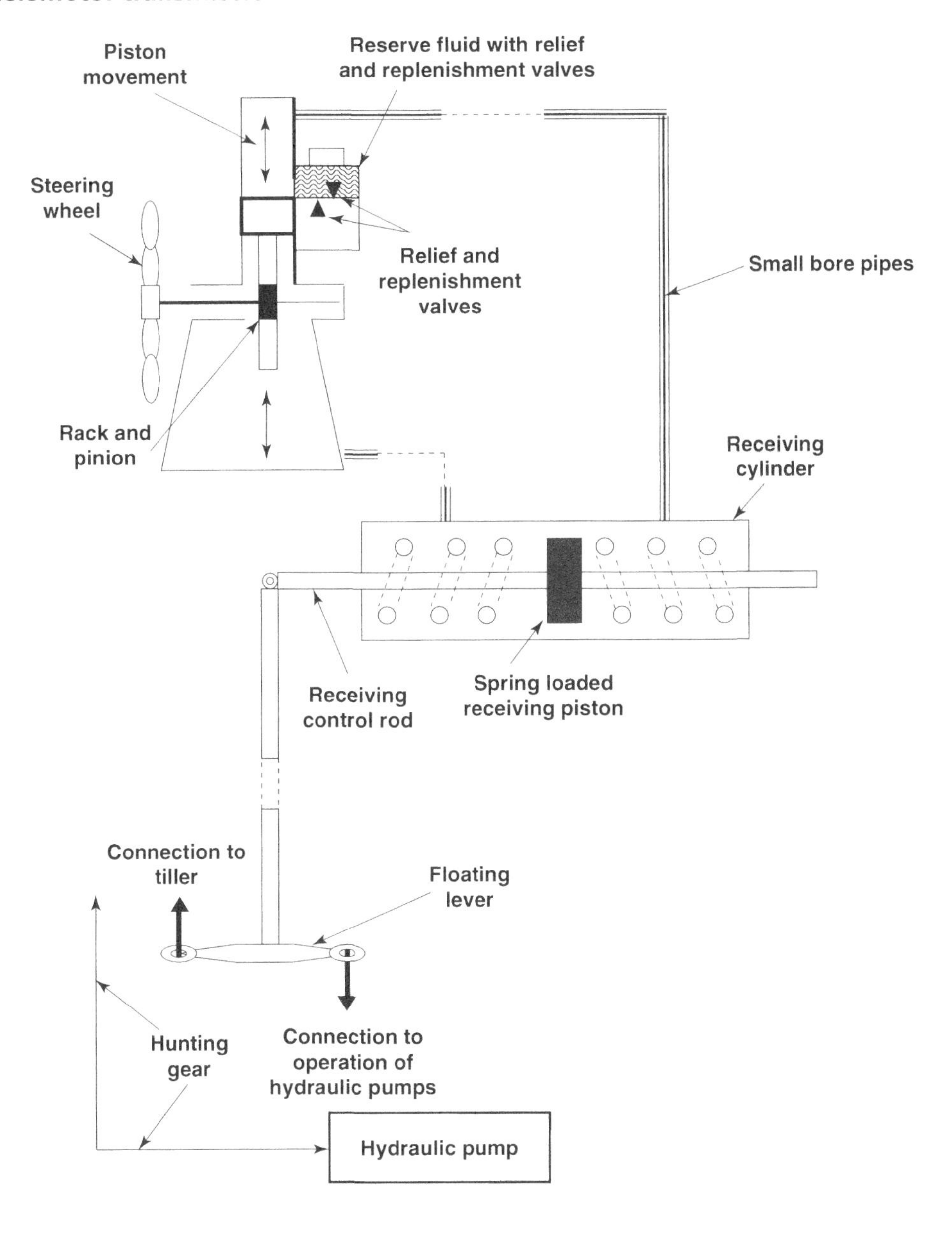

Control system for electro-hydraulic steering gear with telemotor transmission

Hydraulic, telemotor transmission, steering gear consists of five main elements:

a) Telemotor Transmitter
b) Telemotor Receiver
c) Steering motor to activate variable delivery pump*
d) The control element
e) The hydraulic rams.

The steering wheel is geared to two rams which work in cylinders interconnected by piping which corresponds to piping in the telemotor receiver. As the steering wheel is turned, a rack and pinion system engages with the piston and displaces the fluid which then flows through the connecting pipes causing a similar movement at the receiving rams. Linkage to the steering motors activates a variable delivery pump. The function of the pump is to deliver hydraulic fluid (mineral oil) under pressure from one cylinder to the other. This action causes the rams to move, pushing the tiller over to port or starboard as desired.

Telemotor fluid

It should be realized that the fluid in telemotor systems is specific and will be either a 50/50 mixture of glycerine and water having a freezing point of −23°C (−9°F). Alternatives by way of special telemotor oils are obtainable, but oils with low pour points have too much viscosity and would make the wheel operation heavy to manoeuvre.

Rapson slide two ram hydraulic steering system

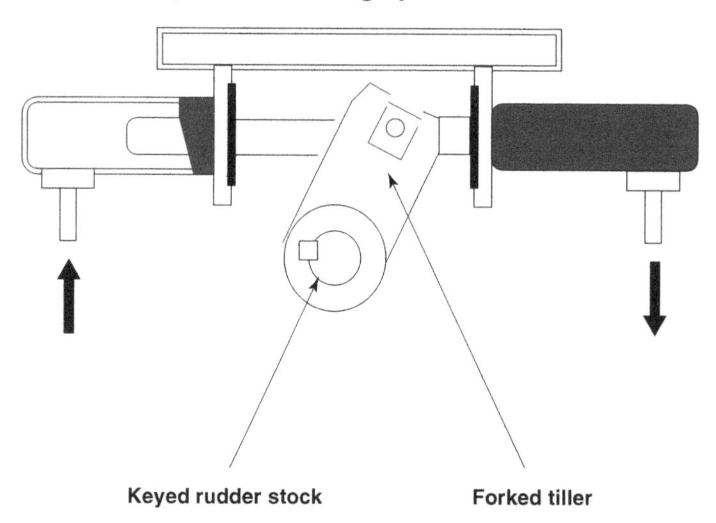

* The hydraulic pump is usually a rotary displacement type driven by an electric motor.

The Rapson Slide mechanism is commonly fitted to two and four ram steering gear systems employing a crosshead arrangement linking the forked tiller. Four cylinders are used for greater power and redundancy, incorporating duplicate pump units.

Steering gear operations

Steering methods have changed considerably over the years, but the basic function has remained – move the rudder to change the movement of the ship's head. Even this basic function has had effective competition from the rotatable thrusters, controllable water jets and the more recent steerable pod technology.

However, new concepts have not yet completely dominated the steerage of ships and many vessels are still fitted with conventional rudder movement in order to control the vessel's heading.

Four ram electro-hydraulic steering system

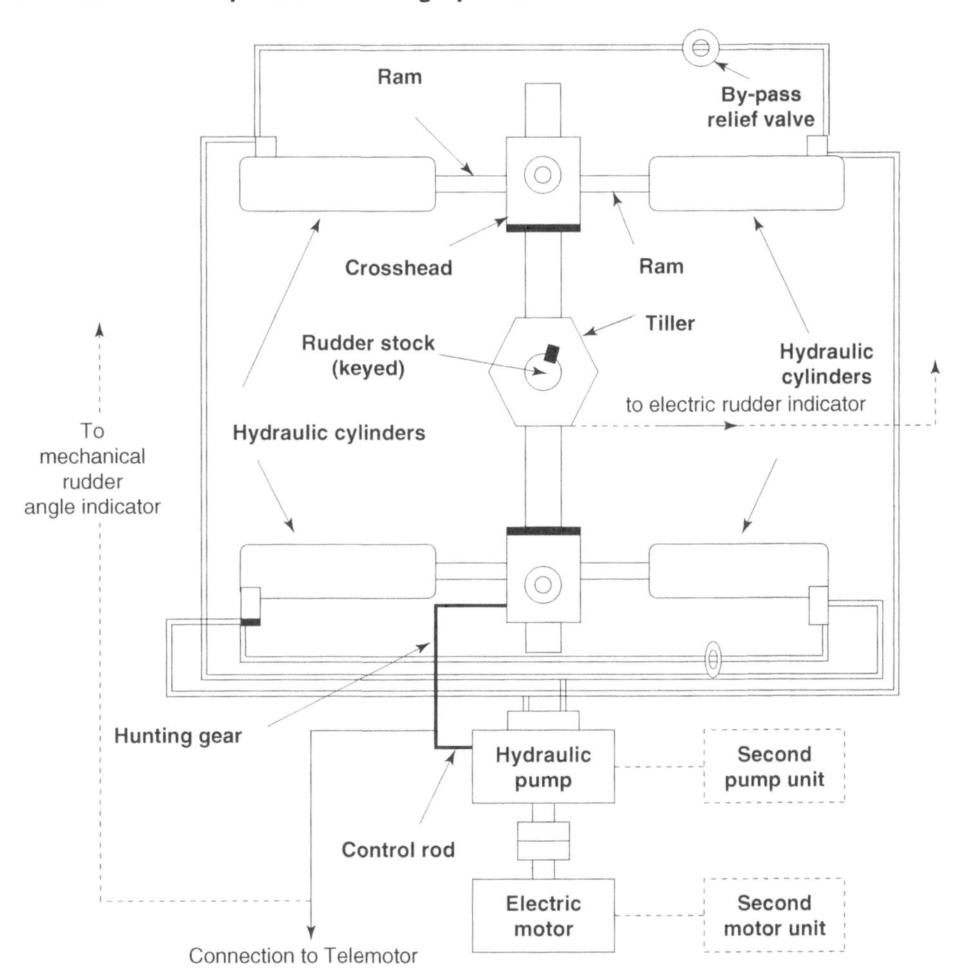

Electric steering gear

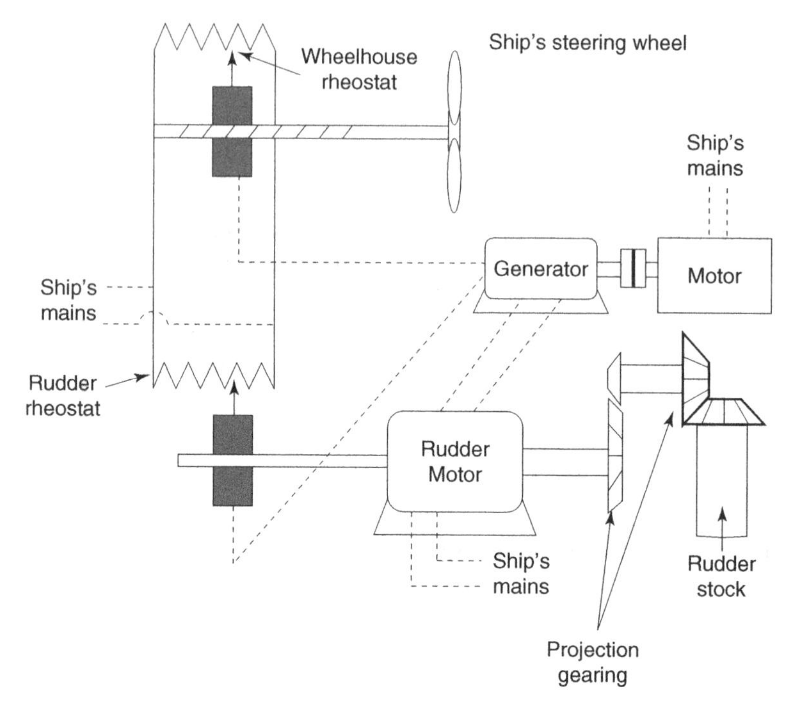

Electric steering gear (based on the Ward-Leonard system).

This system is based on the principle of a Wheatstone Bridge and contains a motor/generator set, which is continuously running while the vessel is at sea. As the steering wheel is turned, the contact on the wheelhouse rheostat is offset and a potential difference will exist. This generates a voltage which causes current flow to the field coils of the generator. The generator then supplies power to the rudder motor, causing the rudder to rotate. The speed of the rudder motor will vary with the voltage supplied to it by the generator and the voltage supplied will be directly related to the potential difference within the field.

As the rudder moves the desired amount, the rudder rheostat will also be caused to move to a coincident position, giving a zero voltage across the field. The rudder motor will then stop and the system comes to rest because the balance of the resistance bridge has been restored. The contact movement of the rudder rheostat acts very similar to the 'hunting gear' on electro-hydraulic steering gear. This system provides sensitive control, faster response and a high torque.

Maintenance of electric steering gears

Although all electric steering systems have shown themselves to be reliable, they do require some standard maintenance checks. Attention to the renewal of the sliding contactors of the rheostats and the contact fingers of the single motor telemotors will be required to be renewed periodically to prevent wear down. The tension springs holding the contact surfaces together may also experience a lessening of tension and may need to be replaced over time.

The higher costs of electrical installations have influenced the reduced numbers of these units being fitted in new tonnage. They have to some extent been superseded by the equally reliable electro-hydraulic steering gears, which have proved more popular with new building.

Rotary Vane steering

Rotary Vane steering is a compact steering unit which is situated on top of the rudder stock. A rotor is 'keyed' onto the stock and the whole is encased by a steel casing known as a 'stator'. The concept allows follow-up and non-follow-up modes to operate with either electric or hydraulic transmission systems.

Model variations allow rudder angles of 2 × 35°, or 2 × 60° with options of up to 90°. The system tends to act as a self-lubricating rudder carrier, as well as generating the turning movement to the rudder. This is achieved by oil being delivered under pressure to one side of the blades of the rotor. With the rotor being 'keyed' to the rudder stock, when the rotor is caused to turn, so does the stock.

Clearly, the direction of turn will be effected by the direction of the pressurized oil affecting the rotor blades. Therefore, in theory, the rotor and stock can turn only one way, namely in the direction of the pressurized oil. However, if the directional flow of the oil is reversed, by reversing the rotation of the oil pump, then the rotor will also be caused to turn in the opposite direction. This pump reversal from one direction to another provides the necessary directional oil flow to cause movement to port and starboard. The oil under pressure is kept contained within the unit by the stator. The stator is dynamically sealed and is leak free, generally providing an effective, alternative steering mechanism within the created pressure chambers.

Rotary Vane steering operation

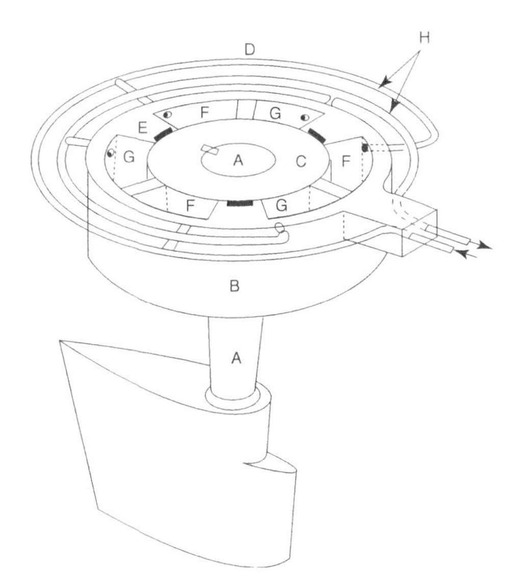

Operation of Rotary Vane steering.

A fixed stator (the outer casing of the steering unit 'B') encompasses a rotor 'C' with fixed vanes attached. The rotor is keyed onto the rudder stock 'A' in such a manner, that when the rotor is turned, the rudder stock, and subsequently the rudder plate, is turned. The system operates with hydraulic oil being delivered under pressure to the chamber 'G' and released out of chamber 'F'.

In order to turn the rudder in an opposite direction the oil pump is given a reverse flow direction, so causing the rotor to move in the opposite direction. This two-way movement can be associated with port and starboard movement by the rudder. The oil pressure acting on the vanes of the inner rotor, generate positive directional movement, depending on the direction of the oil flow.

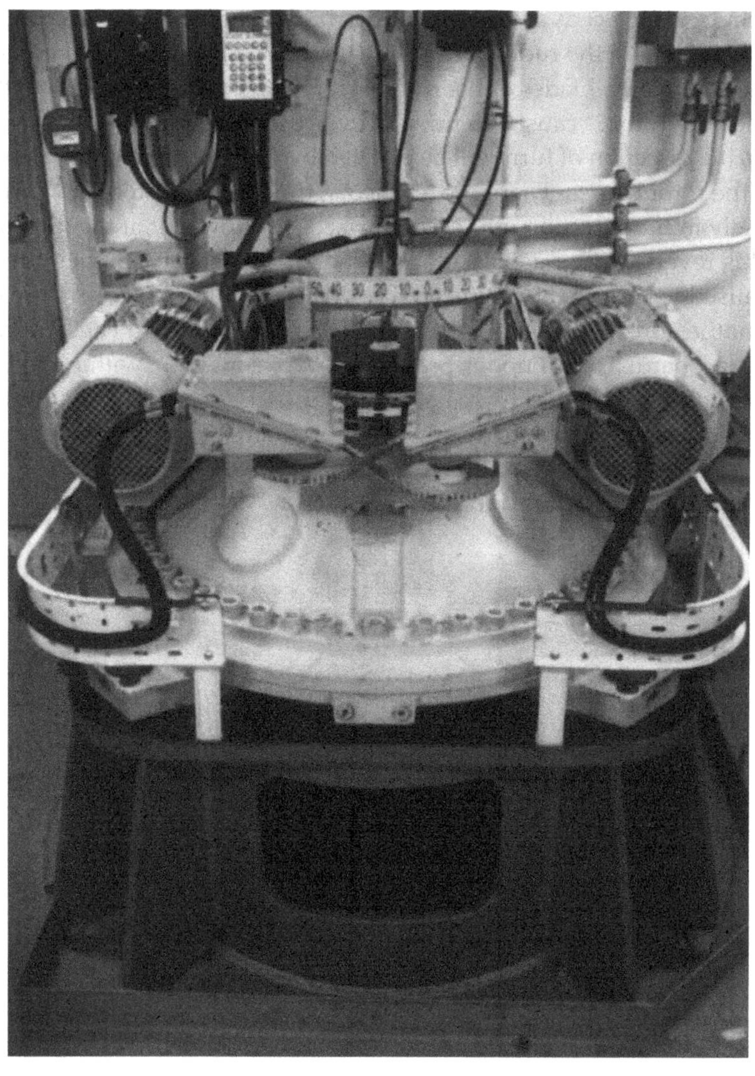

Rotary Vane steering unit. An example of the stator case mounting for a rotary vane steering gear, situated directly over the rudder stock.

Open loop control system for steering operations

If the helmsman is trying to steer a straight course, then a comparison between the Measured Value (MV) of the ship's head and the Desired Value (DV) of the intended course must be made. If these two values differ, then an error exists and the helmsman will apply corrective action by turning the ship's wheel (Manual Steering); the action of the helm being made opposite to the direction of the error.

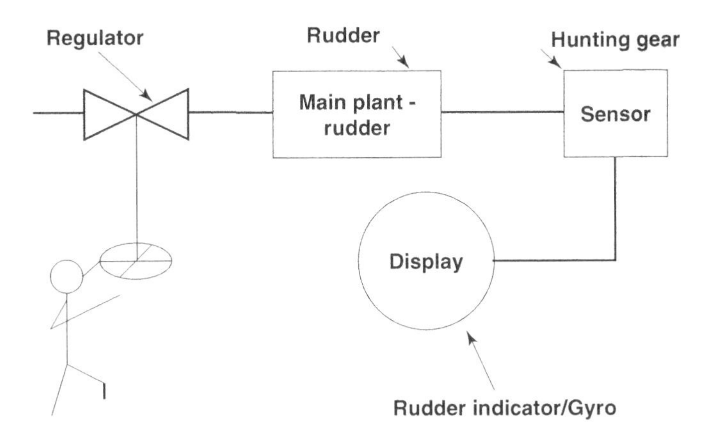

This system does not ascertain the error nor will it use the error to initiate corrective action. Once the helmsman is introduced he or she carries out both tasks of ascertaining the error and applying the corrective action.

Control of transmission (closed loop control)

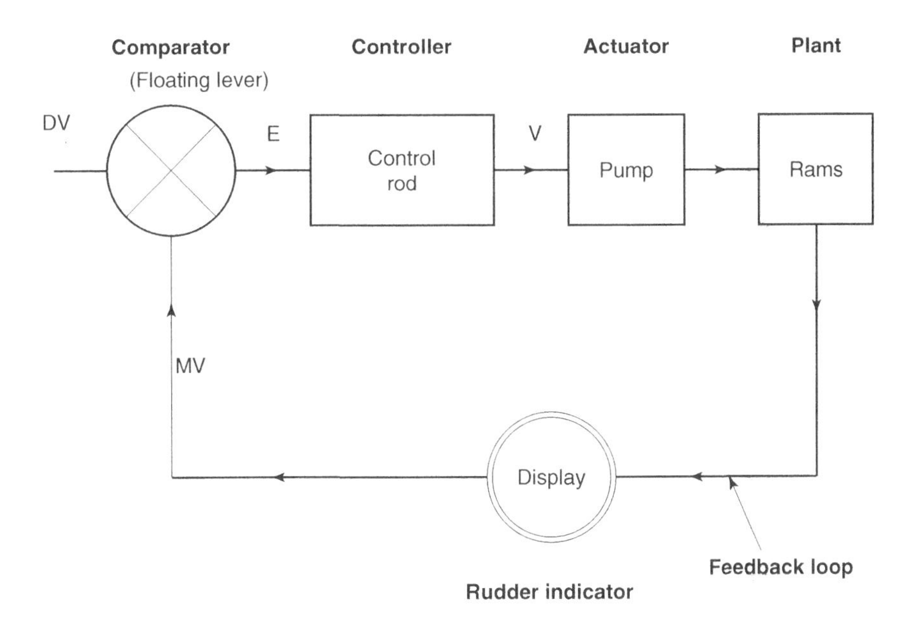

With automatic, closed loop control the comparison between the Desired Value (DV) and the Measured Value (MV) is made by a comparator within the system itself. The output from the comparator – where DV and MV are not the same – is an Error signal (E), which is passed to the controller. This amplifies the error signal and outputs a power signal (V), which can be used to apply corrective action by causing the pump and rams to be moved.

A feedback system that incorporates a rudder indicator on the bridge, displays the action and movement of the rudder, while the hunting lever moves between the rudder stock and the control rod for the pump. This effectively switches off the power when the rams come to rest. Actual movement of the rams is transmitted by means of a control rod to a floating lever, the inner end of which acts initially as its fulcrum which, in turn, is secured to the tiller by a rocker arm. Any movement of the control rod is transmitted from the opposite end of the floating link, through gearing to operate the valves of the hydraulic pump units, causing rudder movement.

Automatic steering operations

The bridge control unit for the 'Automatic Pilot' system will contain four main elements, namely: the on/off switch, a mode selector switch for AUTO, WHEEL or TILLER control, a rudder angle indicator, together with a Gyro compass repeater. A further rudder indicator is usually supplied, which receives its input from the rudder translator as feedback.

NB. The gyro repeater will generally show a ship's head outline to indicate the ship's head, geared to the gyro repeater on an engraved compass card. A second ship's head outline will be incorporated, probably on a second transparent card, which can be turned in the centre to set the desired course value. Differing models have differing limits, but a realistic value is considered as a 45° limit.

Adjusting controls are featured with each model and usually include the following:

Balance control – A control which is adjusted prior to sailing to balance the amplifiers so that the Port and Starboard relays will be activated when the ship's head has swung an equal number of degrees to port or starboard. Once the correct balance has been found, the inner scale is turned until the 'zero mark' is referenced to a line up marking. The scale is then fixed and serves to reference other controlling elements.

Permanent helm – A control which allows for weather helm. It is adjusted to alter the bias on both valves so that one will be activated before the other for an equal error in the ship's head.

Sheering – This control alters the grid bias on the relay valves adjusting their sensitivity. The larger the sheering setting, the greater the angle the ship's head will be allowed to swing through, before the amplifier becomes unbalanced.

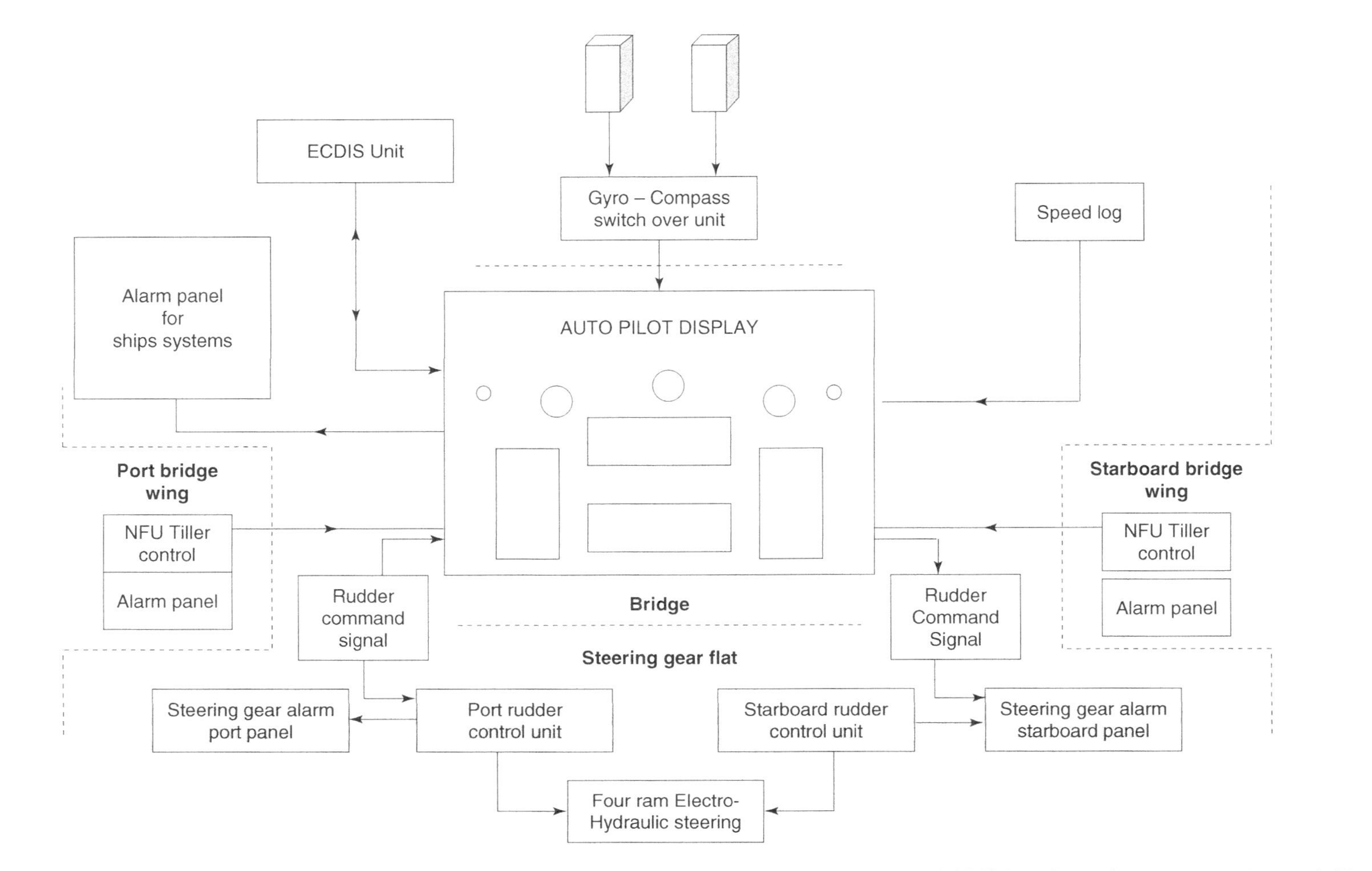

Integrated navigation and steering system for bridge/steering flat. Integrated navigation bridge with ECDIS interfaces for course and speed. The ECDIS would also have an ARPA overlay and depth input.

Damping – An additional control which adjusts the sensitivity by varying the time interval between the instant the ship's head moves off course and the instant when helm is applied. The interval varies between immediate action, a 6 to 8 second delay or a 12 to 16 second delay.

Rudder – This control has several positions, each of which switch in resistors in series along the output from the helm ordered potentiometer to the valve positions. The potential difference across the resistors requires a greater movement of the wiper to restore the balance.

Counter rudder – When a vessel is returned back onto her course, the momentum usually causes an over shoot of the desired heading. This counter rudder control corrects this over shoot: when the vessel is off course and steady; correcting helm is reduced; when the vessel returns towards her course, the counter helm has already been applied and will then be progressively reduced, i.e. the rudder will be returned amidships.

Rudder limit – Is an on or off control. It sets a variable limit in actual degrees on the amount of rudder to be used by the auto-pilot, providing a controlled rate of turn when altering the ship's course.

Phantom rudder (electronic rudder position unit) – Conventional feedback loops for steering gears are fitted with rudder translators. These suffer from the disadvantage that they operate with a time lag. In an operation where the feedback signal orders 'stop', the steering gear and therefore the rudder over shoot. This action can cause instability in the steering. An integrator within the unit will compensate for the delay between the steering gear action and the actual movement of the rudder. It prevents the overshoot and enables the dead band of the rudder to be kept to a minimum to provide more precise automatic steering.

***Note:** Automatic Pilot units and the associated controls vary with manufacturers' models. The more modern versions tend to have less operator input requirements than older models. This is not to say that the same elements are not being compensated for, but designs have developed to incorporate automatic settings for various stabilising effects of the ship's head in a variety of weather/stream conditions.*

Remote controlling station. An enclosed bridge wing control station having duel controls for CPPs, bow thrust and numerous indicators for displaying operational data.

Controlling the hardware

The navigation bridge is the established co-ordination centre for manoeuvring the vessel. The ship may have remote stations, like bridge wings or steering flats where secondary or emergency controls could be employed. However, most vessels have the hub of navigation as the bridge. Direct links from the bridge and/or remote control stations tend to be linked by control systems to the engine room, bow thrust rooms, stabilizer compartments, etc. Engine movement orders are passed via the bridge telegraph, rudder movement being sensed and transmitted back via a feedback loop to a helm indicator at the bridge station, with direct communication lines to all essential compartments affecting the manoeuvring of the vessel.

Bridge indicators also provide feedback on main engine's rpm, rate of turn, pitch angle of CPPs, angle of heel (inclinometer) rudder indicator, navigation light status, and watertight integrity of the hull. Additional sensing devices exist on the larger vessel for watertight doors, fire doors, draught indicators, smoke detectors and/or state of tanks.

The bridge is continually manned while the ship is at sea, and performance criteria can be fully monitored. Similarly, engine control rooms are also continually manned when operating with other than unmanned engine rooms.

The Main Engine Control Room employs analogue and digital readouts of sensed elements. Mimic diagrams are employed for pipeline systems. The modern control room includes connections to Voyage Data recorders and CCTV.

Engine control rooms

The position of the machinery control room is usually situated in a central position so as to afford a good overview of the most essential elements of the machinery space. The room itself should be fitted with double glazed toughened glass windows, providing unrestricted viewing and excluding heat, noise and vibration elements from affecting the main machinery.

The control room should be well-ventilated and well-illuminated, being kept at an ambient temperature of about 20 to 25°C and with a relative humidity of between 5 and 60 per cent. The design layout should reflect the operational needs with due attention to the ergonomics. This involves sub-division in consoles of a low overall height where all controls are within easy reach.

Monitoring indicators should read from left to right or from top to bottom and extensive use should be made of mimic diagrams and generally lend to simplicity of use. Data logging systems are now common and provide virtually continuous coverage of all machinery elements from respective sensors, including: pressures, temperatures, flow rates, status changes, level error margins, etc.

The room should have sufficient space to accommodate and cater for several operators who may be expected to be present under any and all working conditions when at sea, or in port, when engaged in routine or emergency situations.

An example of mimic diagram layout showing cargo tank and piping systems.

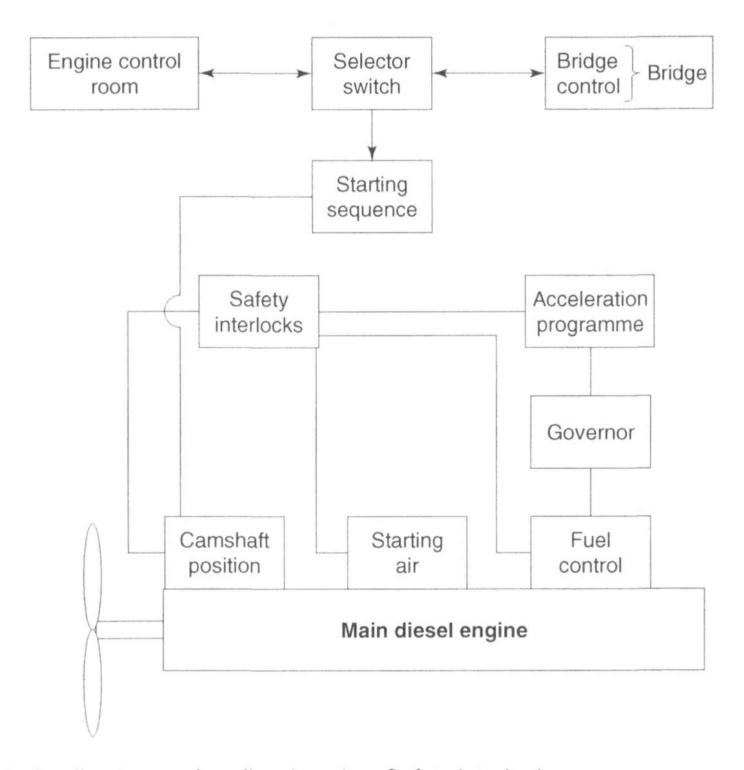

Bridge control – direct reversing diesel engine. Safety interlocks:
a) No start of main engine with turning gear in place
b) No start unless propeller pitch is zero (controllable pitch propellers)
c) No air admitted when engine running
d) No fuel admitted unless correct starting sequence engaged
e) No astern movement unless main engine first stopped
f) Main engine cut out if the governor limits are exceeded.

Monitoring alarm systems for unmanned machinery spaces (UMS)

Unmanned machinery spaces are becoming more common practice throughout the industry. Without doubt they lend to improved efficiency, allowing better maintenance schedules to be established. They are seen as permitting cost-effective use of labour while, at the same time, the use of information technology and solid state systems provides effective monitoring techniques.

Additional expense is incurred by way of instrument duplication to the navigation bridge with some additional transmission systems being necessary. However, the benefits would seem to far outweigh initial outlay costs at the building stage.

Requirements for: bridge control propulsion systems

Bridge instrumentation to include:

1. Propeller speed monitors.
2. Direction of propeller rotation indicators or Pitch Angle position indicators for CPP.
3. Air start pressure gauges for Diesel Engines to show manoeuvring capacity.
4. Emergency 'stop' controls.
5. Audible and visual alarm systems:
   (i) on the bridge
   (ii) inside the engine room environment in the event of power supply failure to bridge systems.

Alarm systems – coverage

a) Machinery faults indicated in the machinery control room.
b) Engineer awareness alarms to faults.
c) Alarm systems designed with self monitoring properties.
d) Power failure supply with indicated alarms.
e) Alarm displays at either the main control station, or a subsidiary station, must have means of identifying the fault.
f) Where the Navigation Officer is the sole watchkeeper, then the bridge is to be made aware of the following:
   (i) Any machinery fault which has occurred
   (ii) That the fault is being attended to
   (iii) When the fault has been rectified.
g) A fire detection alarm system which covers the machinery space and is indicated on the bridge.
h) A bilge level alarm – two independent systems.
i) Automatic light switch on in the event of mains failure.
j) Audio/visual alarms. Where audio is silenced visual alarms remain.
k) In the event of a second fault occurring whilst the first is being attended then the audio alarm is re-activated.
l) Acceptance of an alarm outside the machinery space shall not silence the audible alarm system.
m) Essential machinery must remain capable of manual operation if bridge or auto-control became inoperable.

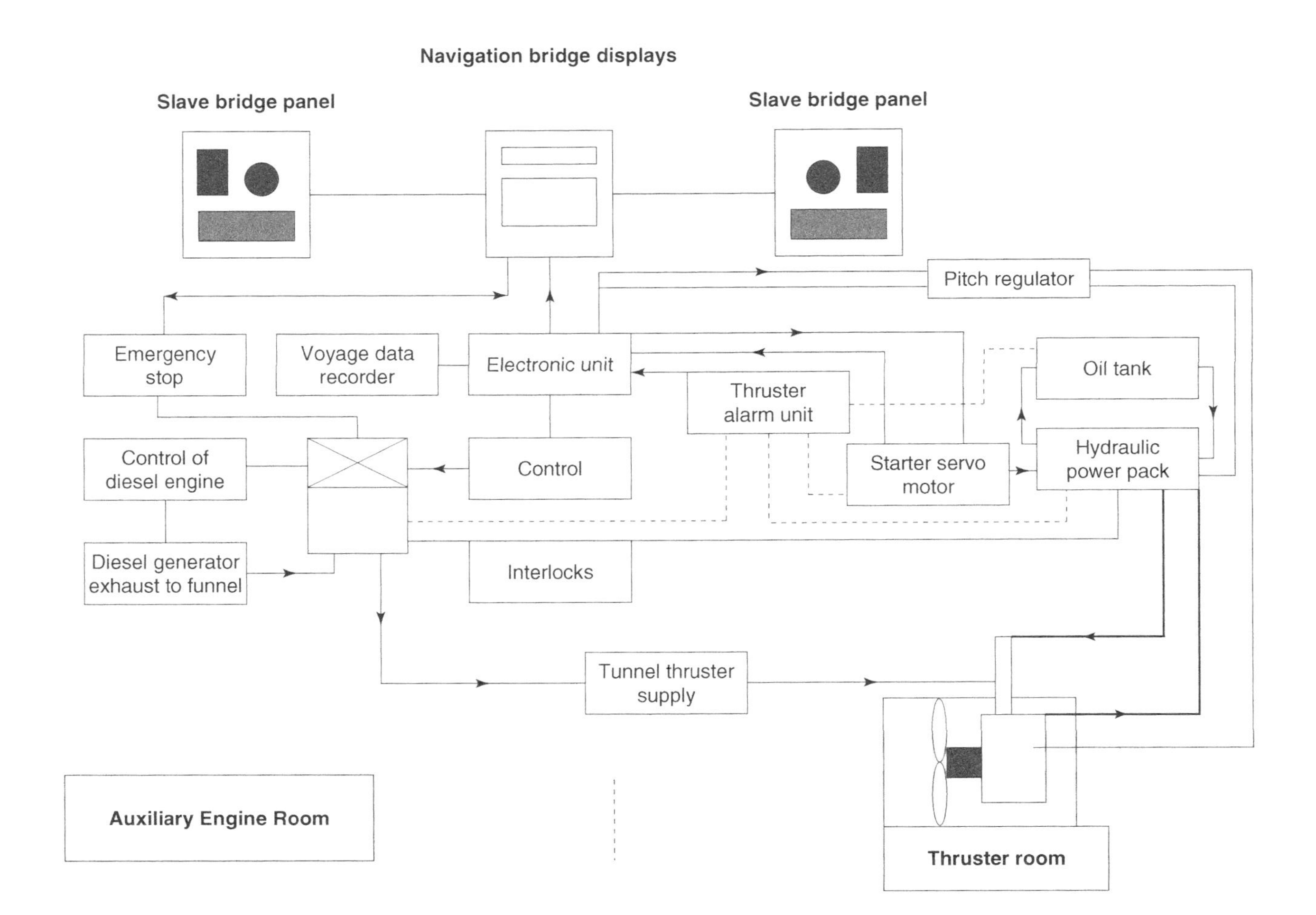

Bridge control of main engine and thruster power.

Appendix B: Danger of interaction

MARINE GUIDANCE NOTE
MGN 199 (M)
Dangers of interaction
Note to Owners, Masters, Pilots and Tug-Masters

This Note supersedes Marine Guidance Notice 18

Summary

This note draws attention to the effects of hydrodynamic interaction on vessel manoeuvrability and describes some incidents which illustrate the dangers.

Key Points:-

- Understand that sudden sheering may occur when passing another vessel at close range
- Appreciate the need to reduce speed in narrow channels
- Be aware of the dangerous effects on tugs when manoeuvring close to larger vessels
- Be aware that unexpected turning moments may result when stopping in shallow, confined basins
- Appreciate the need to make appropriate allowances for squat
- Note the results of laboratory work

1. Hydrodynamic interaction continues to be a major contributory factor in marine casualties and hazardous incidents. Typical situations involve larger vessels overtaking smaller ones in narrow channels where interaction has caused the vessels to collide and, in one case the capsize of the smaller vessel with loss of life.
2. Situations in which hydrodynamic interaction is involved fall into the following categories:-
   (a) Vessels which are attempting to pass one another at very close range. This is usually due to their being confined to a narrow channel.
   (b) Vessels which are manoeuvring in very close company for operational reasons, particularly when the larger vessel has a small underkeel clearance.
   (c) Vessels with a small under-keel clearance which stop rapidly, when approaching an enclosed basin, resulting in unexpected sheering. Included in this category is the reduced effect of accompanying tugs which may sometimes be experienced in these circumstances.

3. Passing vessels

When vessels are passing there are two situations: (i) overtaking and (ii) the head-on encounter.

(i) Overtaking: Interaction is most likely to prove dangerous when two vessels are involved in an overtaking manoeuvre. One possible outcome is that the vessel being overtaken may take a sheer into the path of the other. Another possibility is that when the vessels are abeam of one another the bow of each vessel may turn away from the bow of the other causing the respective sterns to swing towards each other. This may also be accompanied by an overall strong attractive force between the two vessels due to the reduced pressure between the underwater portion of the hulls. There are other possibilities, but the effect of interaction on each vessel during the overtaking manoeuvre will depend on a number of factors including the size of one vessel relative to the other, the smaller of the two vessels feeling the greater effect.

(ii) The head-on encounter: In this situation interaction is less likely to have a dangerous effect as generally the bows of the two vessels will tend to repel each other as they approach. However, this can lead indirectly to a critical situation. It may increase any existing swing and also be complicated by secondary interaction such as bank-rejection from the edge of a channel.

In all cases it is essential to maximise the distance between the two vessels. The watchkeeper on the larger vessel should bear in mind the effect on adjacent smaller vessels and take necessary care when manoeuvring.

4. Interaction in narrow channels

When vessels intend to pass in a narrow channel, whether on the same or opposing courses, it is important that the passing be carried out at a low speed. The speed should be sufficient to maintain control adequately but below maximum for the depth of water so that in an emergency extra power is available to aid the rudder if necessary. If a reduction in speed is required it should be made in good time before the effects of interaction are felt. A low speed will lessen the increase in draught due to squat as well as the sinkage and change of trim caused by interaction itself. Depending upon the dimensions of both the vessel and the channel, speed may have to be restricted. When vessels are approaching each other at this limiting speed interaction effects will be magnified, therefore a further reduction in speed may be necessary. Those in charge of the handling of small vessels should appreciate that more action may be required on their part when passing large vessels which may be severely limited in the action they can take in a narrow channel. Regardless of the relative size of the vessels involved, an overtaking vessel should only commence an overtaking manoeuvre after the vessel to be overtaken has agreed to the manoeuvre.

5. Manoeuvring at close quarters

When vessels are manoeuvring at close quarters for operational reasons, the greatest potential danger exists when there is a large difference in size between the two vessels and is most commonly experienced when a vessel is being attended by a tug. A dangerous situation is most likely when the tug, having been manoeuvring alongside the vessel, moves ahead to the bow to pass or take a towline. Due to changes in drag effect, especially in shallow water, the tug has first to exert appreciably more ahead power than she would use in open water to maintain the same speed and this effect is strongest when she is off the shoulder. At that point hydrodynamic forces also tend to deflect the tug's bow away from the vessel and attract her stern; but as she draws ahead the reverse occurs, the stern being strongly repulsed, and the increased drag largely disappears. There is thus a strong tendency to develop a sheer towards the vessel, and unless the helm (which will have been put towards the vessel to counter the previous effect) is immediately reversed and engine revolutions rapidly reduced, the tug may well drive herself under the vessel's bow. A further effect of interaction arises from the flow around the larger vessel acting on the underbody of the smaller vessel causing a consequent decrease in effective stability, and thus increasing the likelihood of capsize if the vessels come into contact with each other. Since it has been found that the strength of hydrodynamic interaction varies approximately as the square of the speed, this type of manoeuvre should always be carried out at very slow speed. If vessels of dissimilar size are to work in close company at any higher speeds then it is essential that the smaller one keeps clear of the hazardous area off the other's bow.

6. Stopping in shallow basins

A vessel in very shallow water drags a volume of water astern which can be as much as 40% of the displacement. When the vessel stops this entrained water continues moving and when it reaches the vessel's stern it can produce a strong and unexpected turning moment, causing the vessel to begin to sheer unexpectedly. In such circumstances accompanying tugs towing on a short line may sometimes prove to be ineffective. The reason for this is that the tug's thrust is reduced or even cancelled by the proximity of the vessel's hull and small underkeel clearance. This causes the tug's wash to be laterally deflected reducing or even nullifying the thrust. The resultant force on the hull caused by the hydrodynamic action of the deflected flow may also act opposite to the desired direction.

7. Effect on the rudder

It should be noted that in dealing with an interaction situation the control of the vessel depends on the rudder which in turn depends on the flow of water round it. The effectiveness of the rudder is therefore reduced if the engine is stopped, and putting the engine astern when a vessel is moving ahead can render the rudder ineffective at a critical time. In many cases a momentary increase of propeller revolutions when going ahead can materially improve control.

8. General

Situations involving hydrodynamic interaction between vessels vary. In dealing with a particular situation it should be appreciated that when a vessel is moving through the water there is a positive pressure field created at the bow, a smaller positive pressure field at the stern and a negative pressure field amidships. The effects of these pressure fields can be significantly increased where the flow of water round the vessel is influenced by the boundaries of a narrow or shallow channel and by sudden local constrictions (e.g. shoals), by the presence of another vessel or by an increase in vessel speed. An awareness of the nature of the pressure fields round a vessel moving through the water and an appreciation of the effect of speed and the importance of rudder action should enable a vessel handler to foresee the possibility of an interaction situation arising and to be in a better position to deal with it when it does arise. During passage planning depth contours and channel dimensions should be examined to identify areas where interaction may be experienced.

9. Squat

Squat is a serious problem for vessels which have to operate with small underkeel clearances, particularly when in a shallow channel confined by sandbanks or by the sides of a canal or river. The 'Mariners' Handbook' (NP 100) contains further information on squat. The Admiralty Sailing Directions also give specific advice for squat allowances for deep draught vessels in critical areas of the Dover Strait.

Examples of accidents caused by hydrodynamic effects

1. OVERTAKING IN A NARROW CHANNEL This casualty concerns a fully loaded coaster of 500 GT which was being overtaken by a larger cargo vessel of about 13,500 GT. The channel in the area where the casualty occurred was about 150 metres wide and the lateral distance between

the two vessels as the overtaking manoeuvre commenced was about 30 metres. The speeds of the two vessels were initially about 8 and 11 knots respectively. When the stern of the larger vessel was level with the stern of the smaller vessel the speed of the latter vessel was reduced. When the bow of the smaller vessel was level with the midlength point of the larger vessel the bow started to swing towards the larger vessel. The helm of the smaller vessel was put hard to starboard and speed further reduced. The rate of swing to port decreased and the engine was then put to full ahead but a few seconds later the port side of the smaller vessel, in way of the break of the foc'sle head, made contact with the starboard side of the larger vessel. The angle of impact was about 25° and the smaller vessel remained at about this angle to the larger vessel as she first heeled to an angle of about 20° to starboard and shortly afterwards rolled over and capsized, possibly also affected by the large stern wave carried by the larger vessel into which the smaller one entered, beam on, as she dropped back.

2. Manoeuvring with tugs

The second category is illustrated by a casualty involving a 1,600 GT cargo vessel in ballast and a harbour tug which was to assist her to berth. The mean draughts of the vessel and the tug were 3 and 2 metres respectively. The tug was instructed to make fast on the starboard bow as the vessel was proceeding inwards, and to do this she first paralleled her course and then gradually drew ahead so that her towing deck was about 6 metres off, abeam of the vessel's forecastle. The speed of the two vessels was about 4 knots through the water, the vessel manoeuvring at slow speed and the tug, in order to counteract drag, at 3/4 speed. As the towline was being passed the tug took a sheer to port and before this could be countered the two vessels touched, the vessel's stern striking the tug's port quarter. The impact was no more than a bump but even so the tug took an immediate starboard list, and within seconds capsized. One man was drowned.

3. Stopping in a shallow basin

In the third category a VLCC was nearing an oil berth in an enclosed basin which was approached by a narrow channel. The VLCC stopped dead in the water off the berth while tugs made fast fore and aft. An appreciable time after stopping the VLCC began to turn to starboard without making any headway. The efforts of the tugs to prevent the swing proved fruitless and the starboard bow of the tanker struck the oil berth, totally demolishing it.

Results of laboratory work

1. Extensive laboratory work has been carried out on the combined effects of hydrodynamic interaction and shallow water (i.e. depth of water less than about twice the draught) and the following conclusions, which have been borne out by practical experience, are among those reached:
   (a) The effects of interaction (and also of bank suction and rejection) are amplified in shallow water.
   (b) The effectiveness of the rudder is reduced in shallow water, and depends very much on adequate propeller speed when going ahead. The minimum revolutions needed to maintain steerage way may therefore be higher than are required in deep water.
   (c) However, relatively high speeds in very shallow water must be avoided due to the danger of grounding because of squat. An increase in draught of well over 10% has been observed at speeds of about 10 knots, but when speed is reduced squat rapidly diminishes. It has also been found that additional squat due to interaction can occur when two vessels are passing each other.

(d) The transverse thrust of the propeller changes in strength and may even act in the reverse sense to the normal in shallow water.

(e) Vessels may therefore experience quite marked changes in their manoeuvring characteristics as the depth of water under the keel changes. In particular, when the underkeel clearance is very small a marked loss of turning ability is likely.

(f) A large vessel with small underkeel clearance which stops in an enclosed basin can experience strong turning forces caused by the mass of entrained water following it up the approach channel.

(g) The towing power of a tug can be reduced or even cancelled when assisting a larger vessel with small underkeel clearance on a short towline.

Communication and Innovation Branch
Maritime & Coastguard Agency
Spring Place
105 Commercial Road
SOUTHAMPTON
SO15 1EG

Tel 02380 329138
Fax 02380 329204
www.mcga.gov.uk

April 2002
MNA 53/43/001

Safer Lives, Safer Ships, Cleaner Sears

An executive agency of the Department for Transport, Local Government and the Regions

Appendix C: The hardware of manoeuvring ships

Introduction; rudder types and stern arrangements; bow thrust units; stabilizers.

Introduction

The hardware of ship handling continues to move ahead, seemingly on a daily basis. Among them, new concepts are changing, along with hull forms; the wig craft have overtaken high speed craft, at a mere +40 knots; equipment, like pod propulsion units, are superseding yesterday's propellers, just as controllable pitch propellers leapt over the conventional fixed pitch type; bow thrust power is now fitted as multiple units, often accompanied by stern thrust units.

Shipping requirements have been driven by economics and in particular fuel prices. Any innovation that can visibly be cost-effective is being entered into the new build arena. Where manoeuvring aids can save the cost of chartering tugs, then such an inclusion at the building stage is more economical than carrying out a retro-fit at a later date.

Of course, the industry has been profit driven since the days of sail. New ideas, especially those that could involve improved fuel economy, like the addition of ducting to propellers, become attractive to owners. Stabilizers, especially for Roll On–Roll Off traffic, is a clear example, where less rolling means reduced cargo damage claims, so retaining insurance premiums at acceptable levels.

Specific sectors of the industry have seized upon innovation, often adopting ideas from shore side industries and incorporating them into the maritime environment. High speed craft, for instance, are employing controllable jet propulsion units and generating highly manoeuvrable and exceptionally fast, ferry transport.

The life blood of the industry has continued to flow through research and development. No more so than with improved marine systems, used for ship handling. The 'Becker' and 'Schilling' rudders have reduced turning circles dramatically with the additions of flaps and rotors. Bearing in mind all the changes in the past, it will be interesting to see the innovations of tomorrow.

Rudder types and stern arrangements

Rudders

With so many rudder types available to the ship owner today, it would be totally inappropriate to discuss one kind of rudder with possibly a 'Rudder Post' and 'Bearing Pintle'. The fact that the majority of rudders have many variations of rudder bearings and rudder securings, together with different transmission methods from the steering consul, tends to make the operational needs different for each particular rudder.

Rudders are now active in nature, having hydrodynamic flaps, or power motors which provide improved positive response to steering orders. Many ships are fitted with the modern 'Schilling rudders' or the popular hanging 'Becker rudder' or the 'Becker King support' variety.

Stern rudder/propeller arrangements

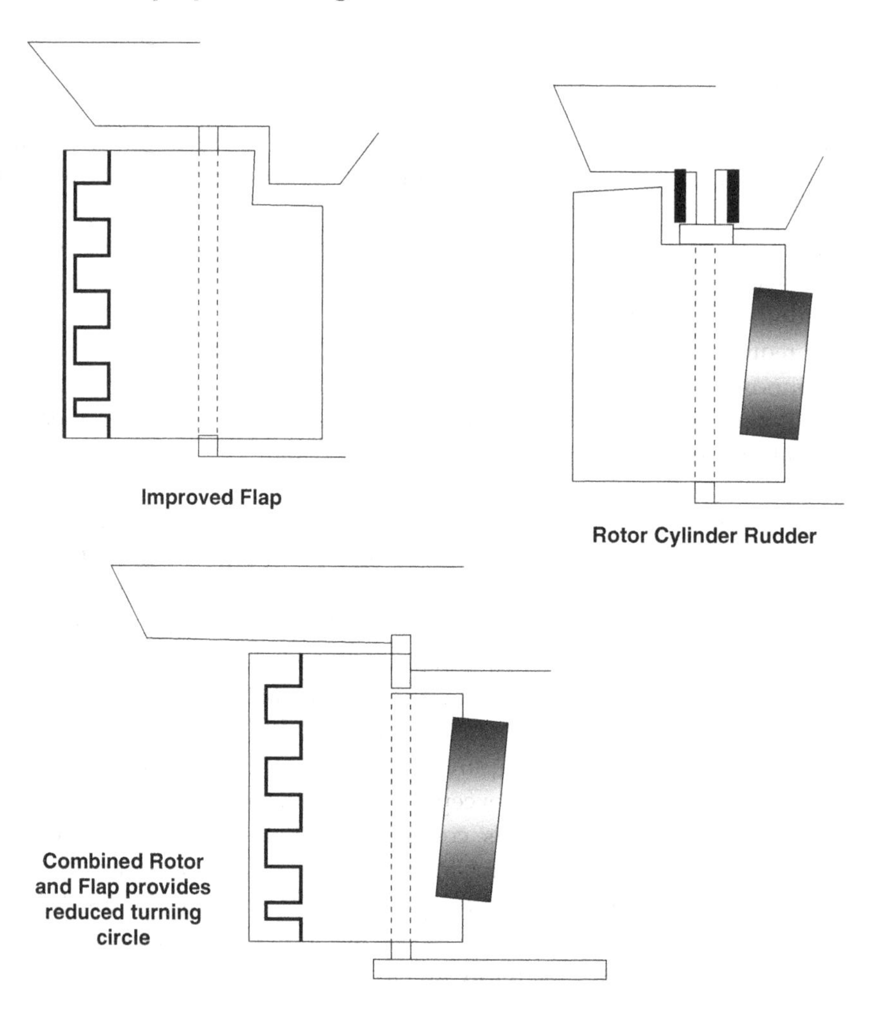

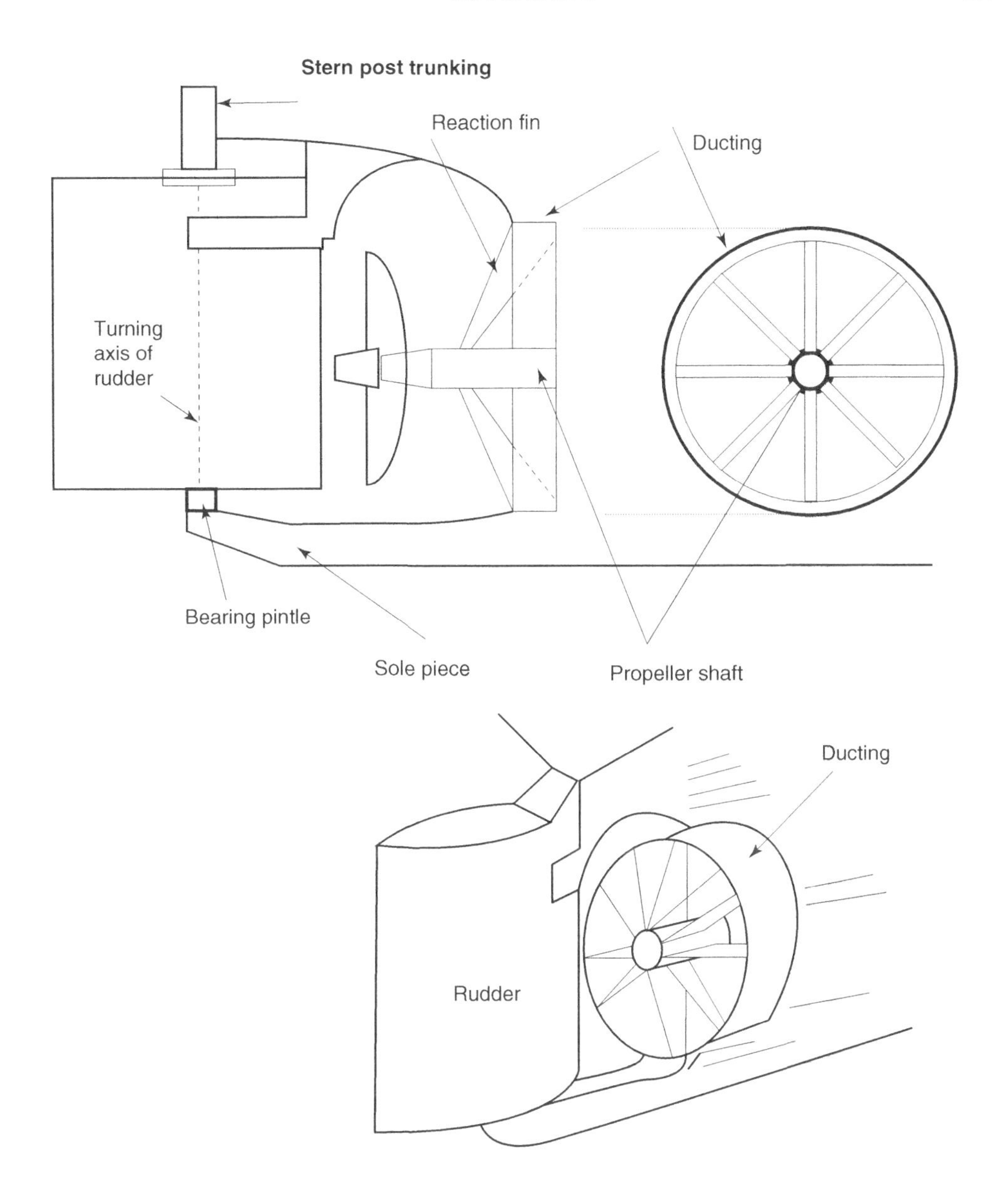

Stern arrangement with propeller ducting and reaction fins. Propeller not included for clarity.

Arguments for ducted propellers

If the ducted propeller is compared with the open propeller, several distinctive aspects are noticed, the most influential probably being the enhanced speeds gained of up to about 0.6 knots. This can clearly be translated to a fuel burn/power saving over an example voyage. Such a saving could justify the initial expense of fitting the ducting arrangement at the building stage.

Experience has also shown that ducting may reduce vibration effects, particularly where cavitation is a feature. In comparison with an open propeller, vessel vibration about the stern and the 'thrust block' is noted as being suppressed by the installation of ducting. The reduced vibration effects may be as a result of the reduced overall size of the propeller where ducting is employed, as compared to a larger propeller in open design where ducting is not present. An exception to the norm would be where a vessel is a lightship or in ballast when excessive vibration (nearly twice the normal levels) can expect to be experienced. When the vessel is deep laden and fitted with ducting, such a vessel benefits from reduced vibration effects. Assuming the ship is gainfully employed with continuous cargoes, ducted propellers would seem to be a favourable addition.

Obviously with smaller propellers being used in construction (including the spare) production costs are reduced. However, this reduction would be offset by the cost of installing the duct itself. In practice, if a vessel suffers damage in the propeller region, the duct itself may suffer damage. This may be seen to afford some protection to the propeller and the duct could be repaired by regular shipyard methods; whereas a damaged propeller would need to be drawn and possibly recast.

Ducted propellers would seem to be economically viable to the ship owner. However, corrosion on such an additional fitment below the waterline must also be considered in the economic equation. It would also seem to be better to fit at the new build stage than to retro-fit ducting to an existing vessel. Retro-fitting is more labour-intensive and therefore some thought should be given at the design stage as to the needs of the trade and the voyage plans.

Stern arrangements with ducted propellers. Flap rudder aft of the propeller duct on a new build twin screw anchor handling vessel. Twin stern thrusters are seen built into an enhanced skeg. The whole area is fitted with sacrificial anodes because of the use of dissimilar metals employed in the manufacture of the hull and propeller units.

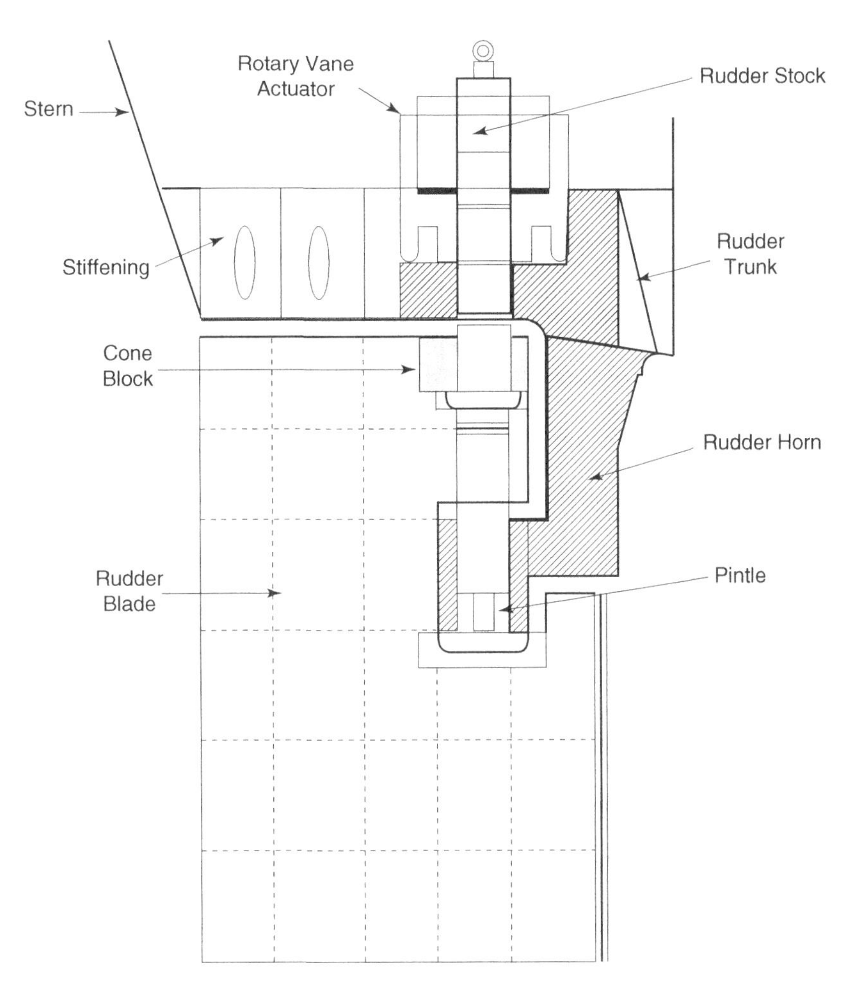

The Mariner rudder.

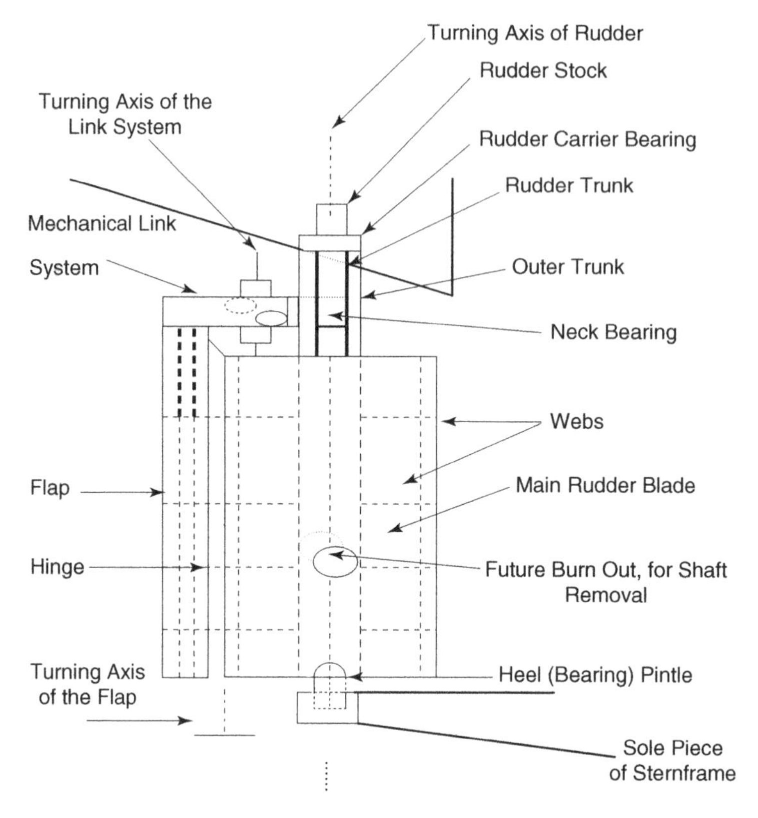

Flap rudder arrangement. General arrangement of 'Becker' standard type flap rudder.

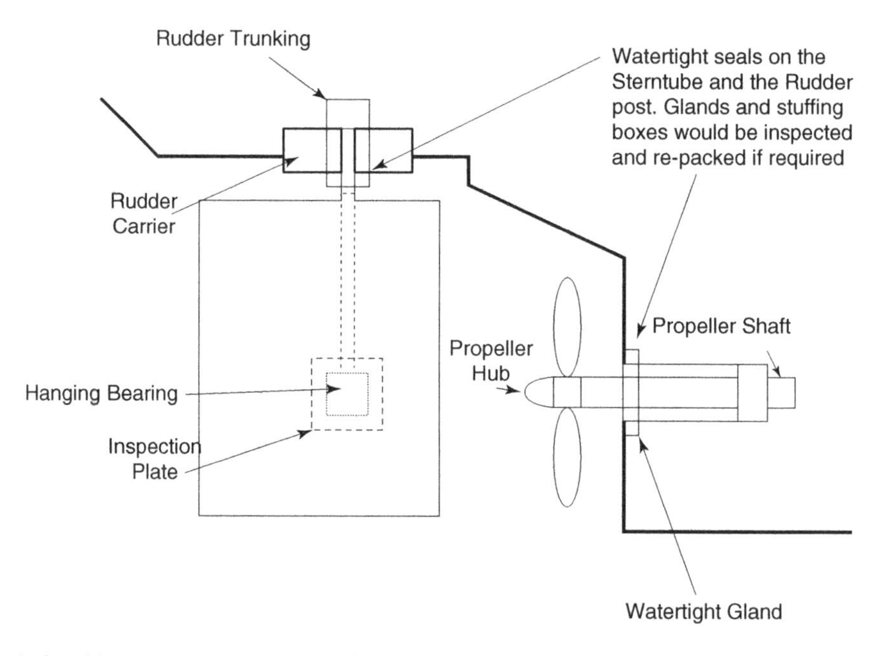

Suspended rudder – stern arrangement. Hanging/suspended rudders.

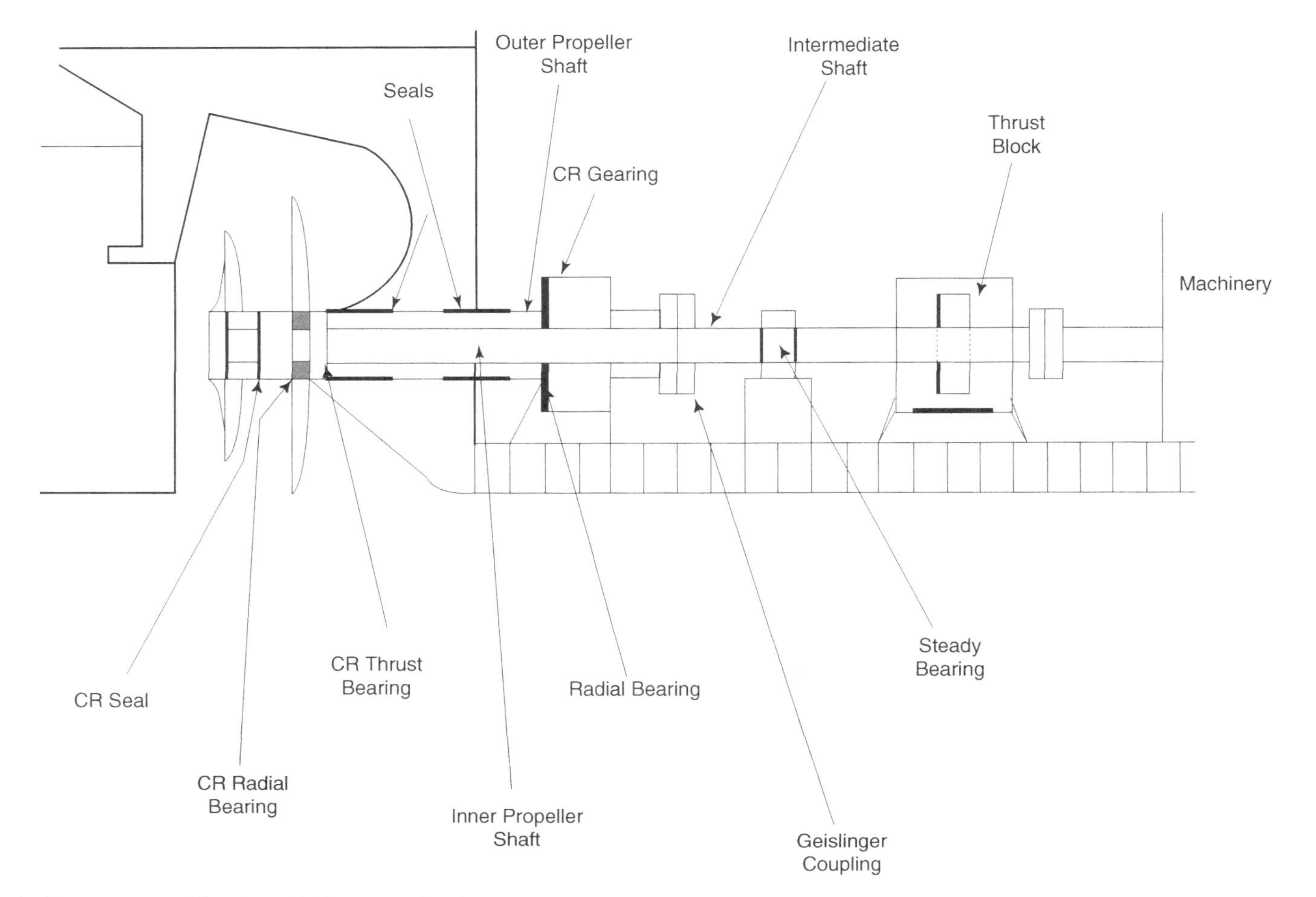

Propeller shaft arrangement (contra-rotating propellers).

Becker King support rudder

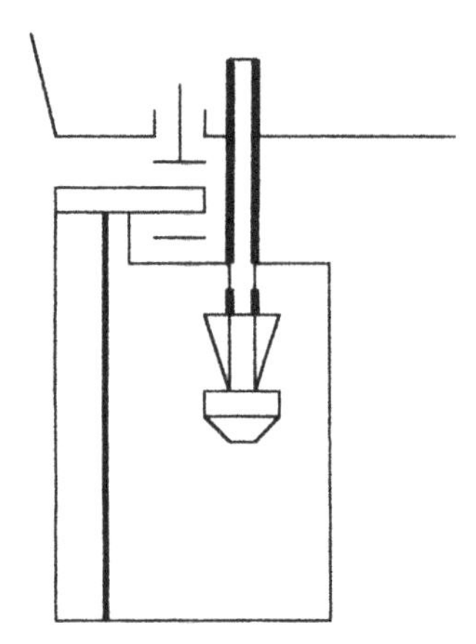

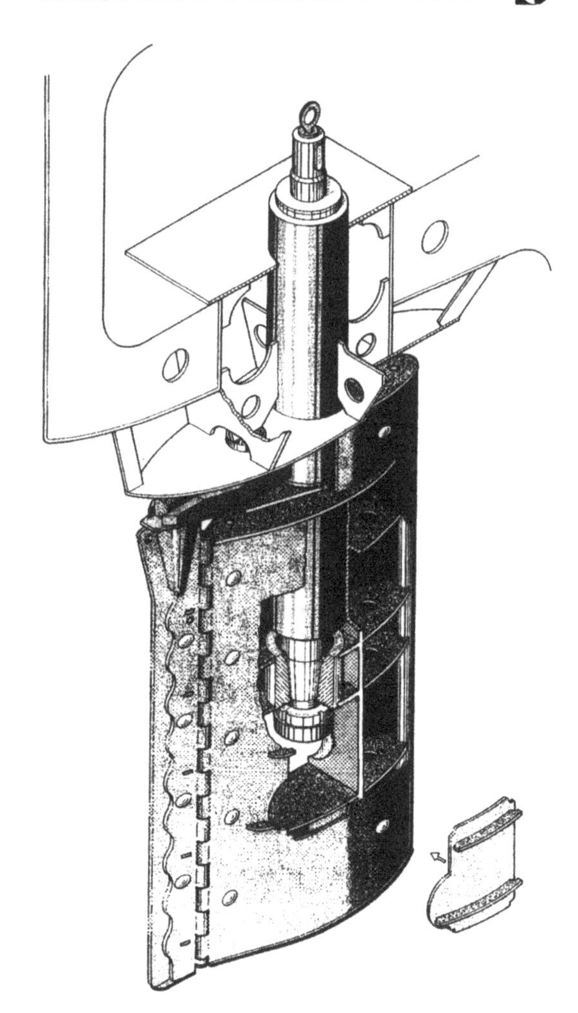

Becker-Ruder support rudder.

These rudders are connected with the ship's hull by means of a rudder trunk which is a cantilever design with an inside bore for location of the rudder stock. The end part of the rudder trunk is provided with an inside bearing for mounting the rudder blade. The exposed, lower part of the rudder stock, is fixed to the rudder body for input of the torque.

Schilling rudders

With the advance of rotors and flaps the turning circle of vessels was able to be considerably reduced. The introduction of the schilling rudder meant that a further major reduction of the turning circle was achieved.

This particular design is a single-piece construction with no moving parts. It has a hydrodynamic shape fitted with end plates, which help to extract slipstream rotational energy. The trailing wedge reduces yaw to provide excellent course stability according to the manufacturers.

When operational, the 70 to 75° helm position takes the full slipstream from the propeller and diverts it at right angles to the hull, eliminating the need for stern thrusters; an ideal property for berthing operations, giving a sideways movement to the vessel.

The rudder design may be a fully hung spade, or a simplex type, with a balance of about 40 per cent. Sizes of rudders vary, but the positioning should take account of the recommended distance between the trailing edge of the propeller and the stern of the vessel, to equal $1\frac{1}{2}$ times the propeller diameter.

Rudder movement is achieved by steering motors supplied by reputable manufacturers, although several fittings have employed rotary vane steering gear with extremely responsive results; torque affecting the stock being similar to conventional rudder designs.

Schilling VecTwin rudders

This system of twin Schilling rudders operated by rotary vane steering is expected to provide improved course stability over and above a conventional rudder. From the point of view of the ship handler, a single joystick control provides comparatively easy manoeuvring, the ship moving in the direction of the joystick movement; the propulsive thrust being proportional to how far the stick is pushed from the 'hover' position.

Again there are no moving parts underwater, so providing reduced maintenance and less wear and tear compared to conventional rudders. A further advantage is that with the rudders in the clam shell position, the stopping distance of the vessel is greatly reduced and the heading is still retained. The arrangement tends to work well with bow thrust in opposition at the fore end, giving the vessel an extremely tight turning circle practically within its own length.

If the turning circle of the vessel with a VecTwin rudder arrangement is considered, the rapid speed reduction caused by the large rudder angles (65/70°) would expect to result in a reduced angle of heel. Also, because the speed is reduced so are the 'advance' and 'transfer' values of the vessel in the turn. This subsequently provides a tighter turning ability than, say, a conventional vessel.

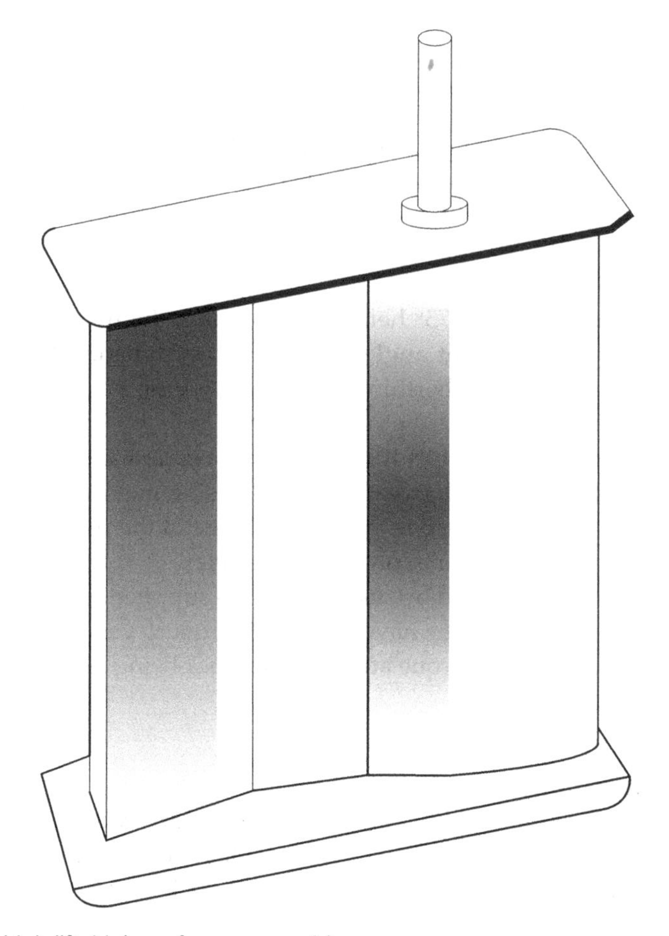

The Schilling, high lift, high performance rudder.

Bow rudders

The bow rudder is a specialist feature and generally not a common fitting other than to ferries and similar vessels that have a need to navigate stern first on a regular basis. Its effectiveness comes from the main propeller's right aft, which sends the wash of water forward passing under the keel of the ship.

The bow form is shaped to accommodate the structure of the bow rudder and frame spacing is reduced in the structural area to provide added strength to accept the rudder activity. The pivot point of the vessel will have a tendency to move towards the bow region to a position of about ⅓ L, measured from the forward perpendicular, when navigating with astern propulsion. The position of the rudder, in conjunction with the new position of the pivot point of the vessel, will usually achieve maximum effect from the arrangement.

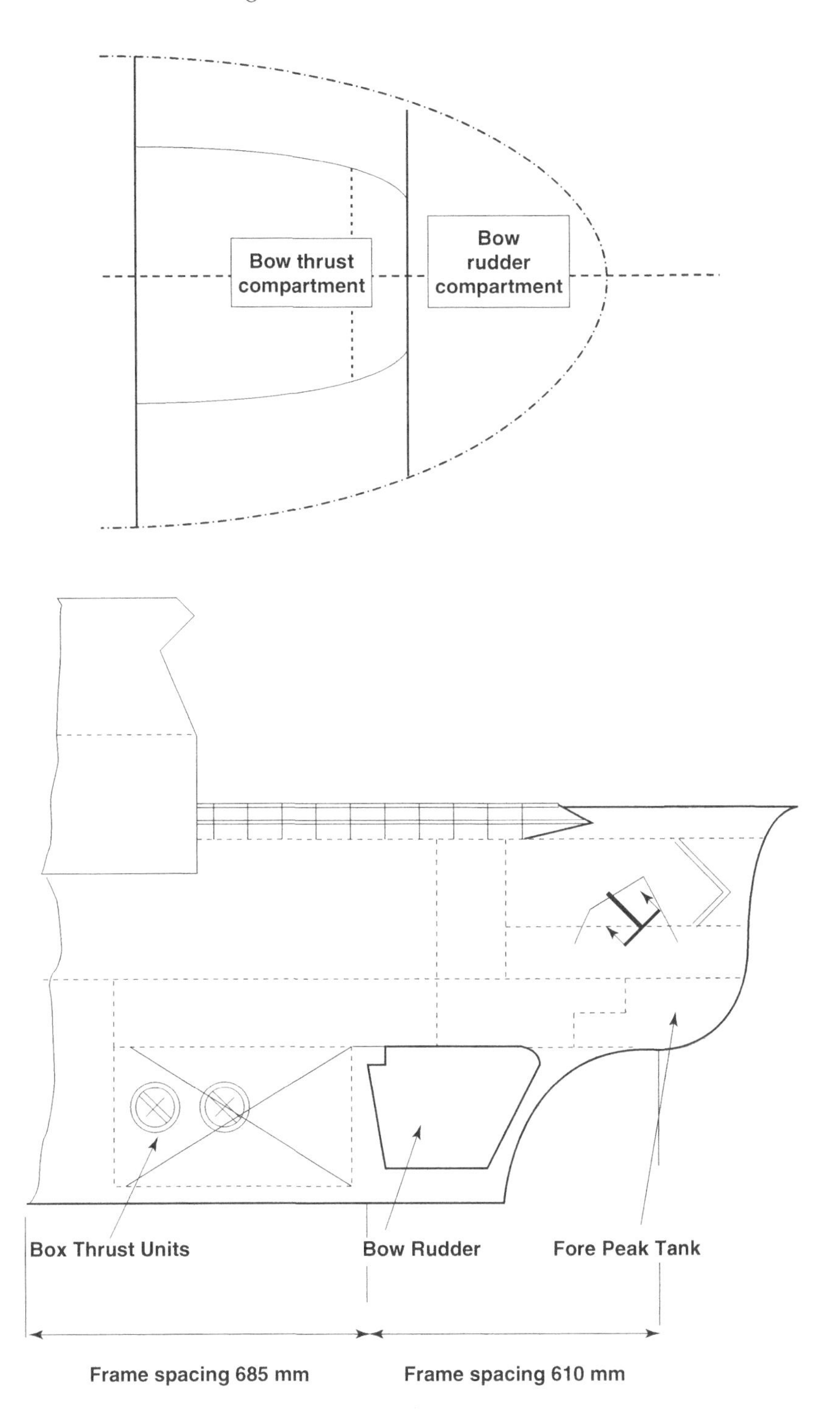

The bulbous bow

A bulbous bow feature is more often fitted to vessels with a large block coefficient; the function being to reduce the water resistance around the bulb and so increase the speed. To understand the principle associated with the expected improved performance, a comparison could be made with a spherical buoy in a strong running tide – as the water passes the buoy, a wave is generated behind the buoy. The bulbous bow is representative of the spherical buoy shape and effectively reduces the bow wave and therefore the vessel experiences less resistance, so obtaining greater speed.

It is important to remember that the bulbous shape increases the wetted surface area, while at the same time provides added forward buoyancy. It has long been felt that, although increasing the wetted surface area will increase hull resistance, the gained advantages of speed outweigh extra frictional resistance affecting the hull.

Bulbous bow constructions can be said to reduce overall resistance of the vessel in a seaway. The shape may improve sea-keeping and provide increased course-keeping abilities. They may have an ice-breaking capability at the upper side of the bulb causing ice sheets to raise and broken ice turning down the sides of the vessel, although the thickness of the ice must be less than that as compared with a formal ice-breaking bow form.

On the downside, anchor positioning with regard to avoiding anchors striking the bulb area, often require additional built-up hawse pipes to allow the anchor to fall clear. The ship's overall length is also generally increased and could impose docking/berthing restrictions. Bow thrust units or echo sounding units may have to be positioned to suit the bulb design and may not result in an optimum user position being possible.

Example of a bulbous bow construction seen exposed in a dry dock situation.

Bow thrust/stern thrust units

Where the needs of the trade dictate, especially where vessels are in and out of port, possibly several times a day, increased manoeuvring aids like thrusters can, in many cases, eliminate the need for expensive tug hire. These units are initially expensive but, if incorporated in a new build from the onset, tend to be valued items for use by the ship handler. Where thrusters are retro-fitted to existing tonnage they become very expensive indeed and will undoubtedly cause internal design change to the vessel.

The design of thruster units vary and top of the range include controllable pitch blade propellers to generate a variable thrust. The more common variety is fitted with fixed pitch blades and a choice option of 50 or 100 per cent thrust power. Many units are fitted with protective grates to prevent sub-surface debris being drawn into the propeller action. The surround is also usually protected by sacrificial anodes in a similar manner to the localized area of any other propeller, where dissimilar metals are present.

A triple bow thrust set exposed in dry dock and undergoing service and maintenance. The upper part of the hull, above the waterline would normally show an indication sign to warn other shipping of the position of the thrusters.

Bow thrust – advantages

Virtually all new large tonnage, especially ferry and passenger vessels, are now being constructed with one, two or even three bow thrust units with or without stern thrusters. Others are being fitted with 'Schilling' or similar style rudders to benefit tight manoeuvres. The clear advantage of using such expensive additions is to reduce the need for tugs during regular docking operations. Without the additional manoeuvring aids, certainly the larger ship would become an expensive commodity to berth and unberth. Also many ship's Masters are now conducting their own ship's pilotage and thereby saving pilotage costs. If this is coupled with removing the need for tug-assistance, for a vessel which might be docking and undocking four or six times per day, the savings are considerable.

Many of the larger ferries are also now being made accessible to the smaller secondary ports because of the skill of the ship handler, with the improved manoeuvring aids. River berths and inland waterways previously did not always permit turning. However, such areas, when navigated with twin CPP and bow thrust units, are now accessible to that larger vessel. 'Stern to' berthing to permit the use of stern ramps has become an operational necessity for ships with inward facing bow berths and outward turning doors, or the bow visor, increased potential docking arrangements, provided the vessel could turn around for the outbound navigation. Bow thrusters permitted and enhanced the tight turns and removed the need for expensive tug use.

These additional aids are costly to install, especially retrospectively. Whereas at the time of new building they are still an expensive addition to the overall building costs, they would be installed as long-term cost-effective equipment; the initial high outlay being offset by the amount of use and saving in additional pilotage fees that would accompany a conventional vessel's docking operations. Neither should the expanding possibilities that may open up to the vessel, during its natural life span, be underestimated. Also, the vessel gains flexibility to operate in additional port options as new port operations come on line.

Conversion costs to existing tonnage are expensive because they often incur high costs in changing already existing designs. Changing and removal followed by fitting new, is always much more expensive than just fitting new from the very start of a ship build. Justification for the change must be strong in the face of high economic pressures affecting a shipping company. The case to install must be seen to deliver cost-effective savings in such areas as pilotage and tug use. Additionally, any vessel with such equipment fitted is clearly seen as a preferred option in the case of a ship sale or a charter option as compared to, say, a vessel without full manoeuvring aids.

On the down side, in addition to the high cost of installation is the fact that any such equipment incorporates maintenance costs and regular dry dock inspections. They are also susceptible to damage from floating debris being drawn into the blade rotation. Generally, they occupy considerable access space aboard the vessel for what is perceived as a small but necessary operational component. Overall the advantages would seem to far outweigh the disadvantages and the ferry sector operators, and especially the ship's Masters, have encompassed bow-thrust units as an essential element of a ship's manoeuvring arsenal.

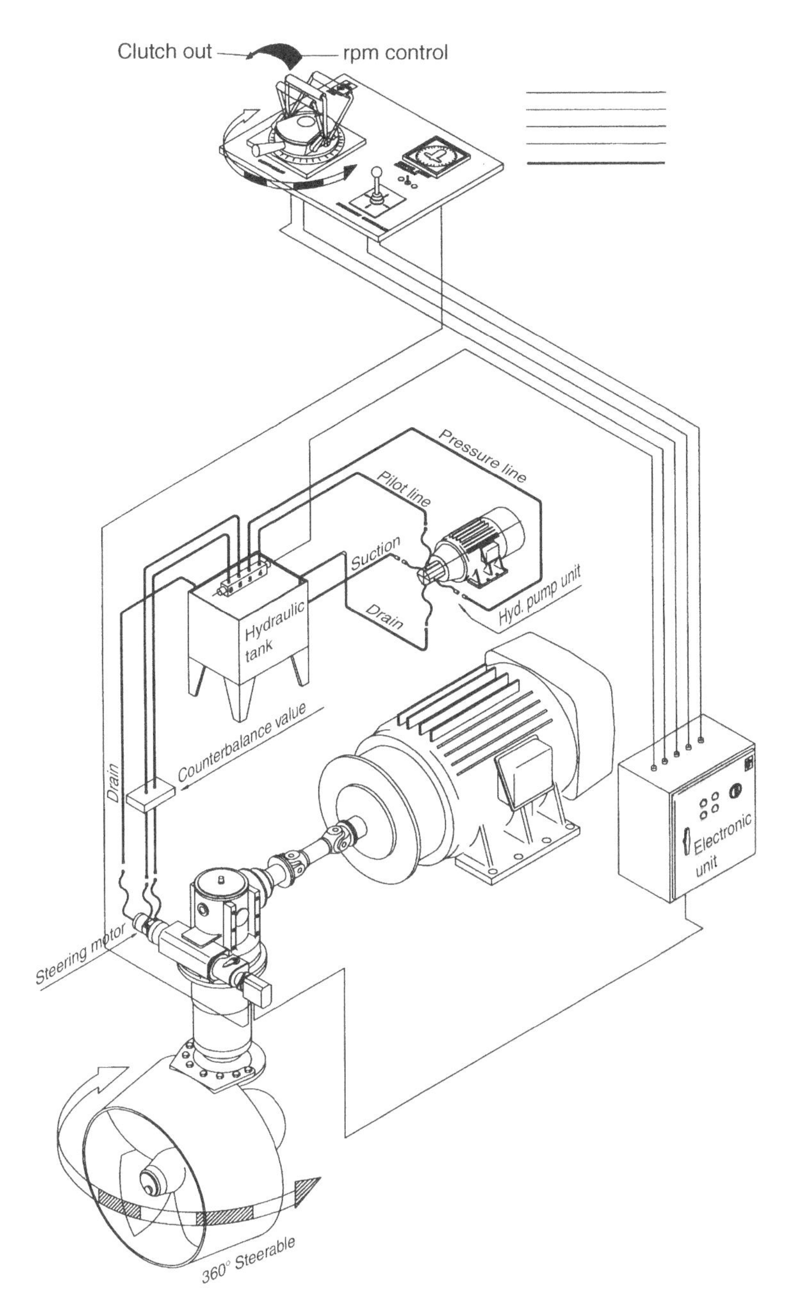

Rota table ducted thruster unit (Azipod operation).

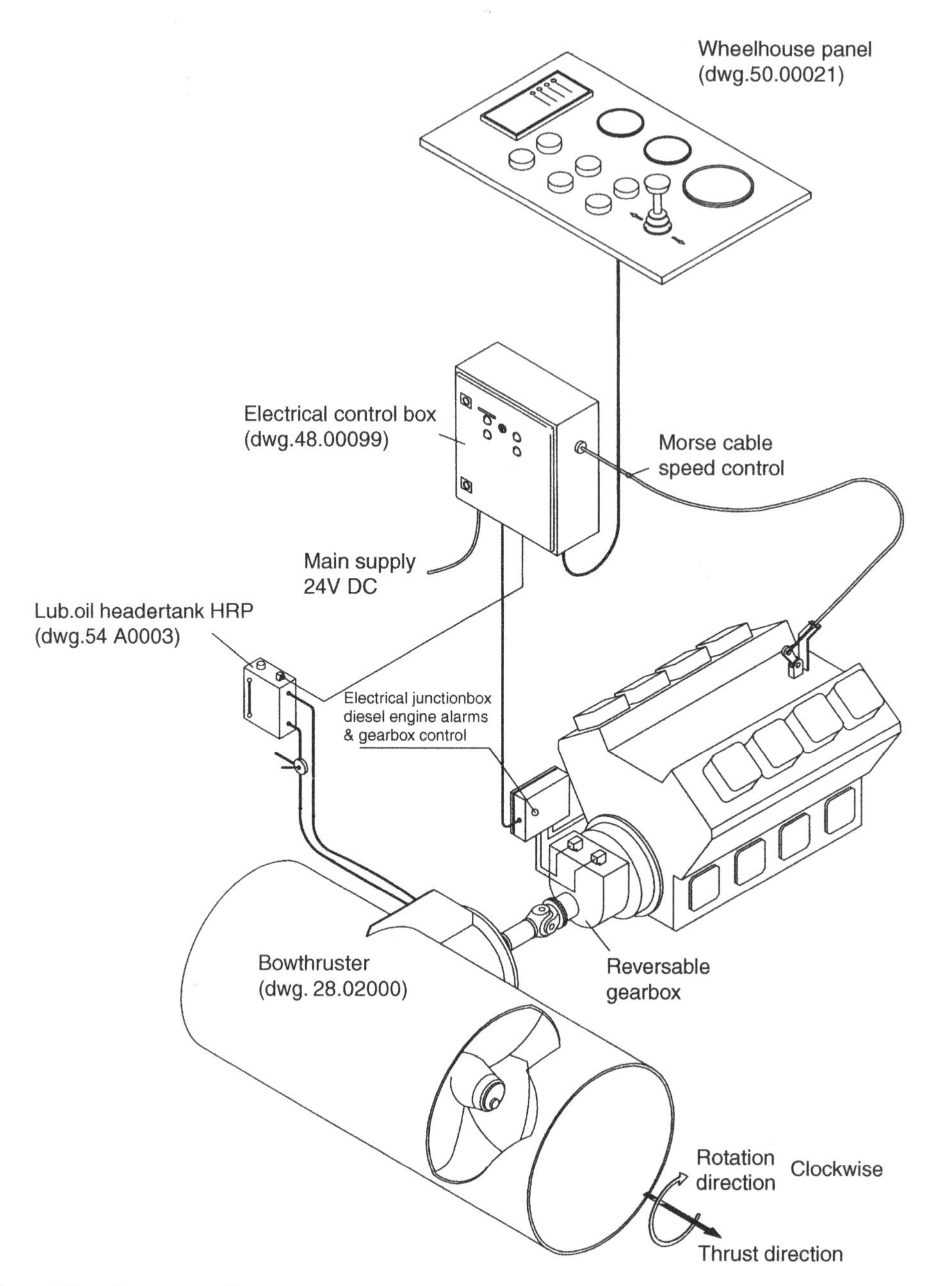

Tunnel thruster – operation.

Retractable Azimuth thruster (Azipod)

Azimuth thrusters have become very popular, with many different types of vessel being fitted with fixed or retractable units. They may also be fitted alongside tunnel thrusters effectively increasing the direction force under the control of the ship handler.

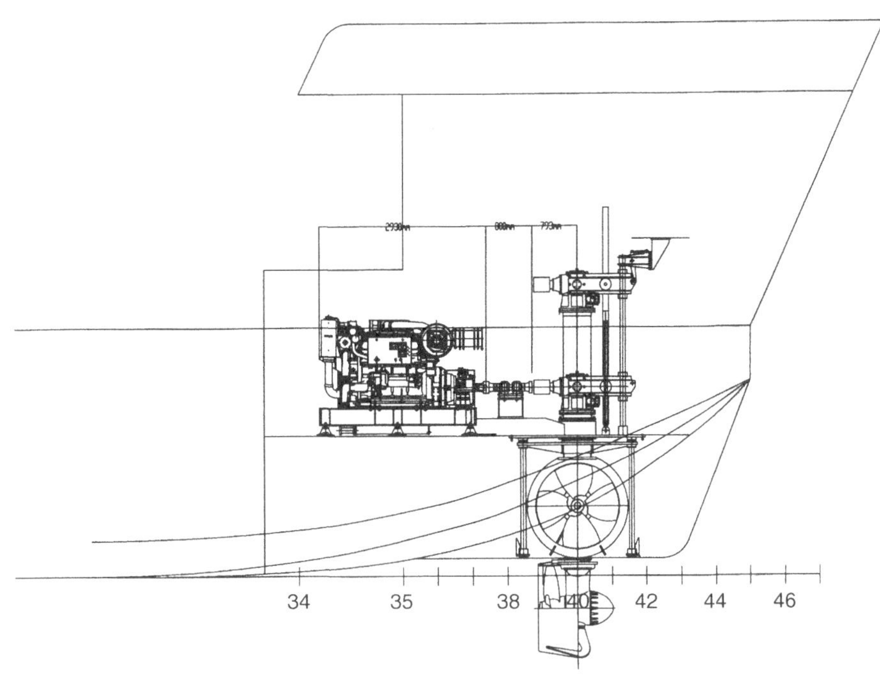

Bow fitting for a retractable azimuth thruster unit.

Although the manoeuvring advantages of azimuth thrusters are clearly beneficial to the ship handler, they do have some disadvantages. The fixed units reduce the underkeel clearance of the vessel while, if the retractable type is fitted these must be deployed to be operational.

They are also classed as an appendage to the vessel and as such would need to be marked on the dry dock plans for the ship. The danger and associated damage incurred would be considerable should fixed azipod units land on blocks as the vessel enters dry dock. Therefore, increased block height would be required in many cases where the retractable units are lowered in dock for maintenance purposes.

Azipod construction

Many types of vessels are now fitted with steerable thruster units, known as 'azipods'. They operate as rotatable thrusters providing not only main propulsion but also steerage to the vessel. They are usually fitted as a multiple feature and are usually ducted but not always. They tend to give high manoeuvrability to a vessel, especially when featured as an addition to main propellers. They are extensively employed with vessels which operate with 'Dynamic Positioning'.

A ducted azipod, featured on the outer hull port side, of a twin rudder/twin propeller vessel.

Stabilizers

Not every vessel is fitted with them but specific vessel types need them, e.g. Ro–Ro vessels. Preventing the vessel rolling in a seaway will greatly reduce the possibility of vehicle cargo shifting and subsequently reduce cargo damage claims and often ship damage claims.

Probably the most effective are fin stabilizers, of which there are two types:

a) Fixed fins or
b) Retractable active fins.

Additionally, many vessels like Ro–Ro's and container ships use stabilizing tanks. These provide a dual function of not only stabilizing the vessel at sea but also keeping the vessel upright when working respective cargoes. Container guides have to be kept vertical with the vessel on an even keel, while the vehicle ramps for Roll On–Roll Off activity needs the even keel condition to retain a flush landing of the ramp to the shore, to permit vehicle movement.

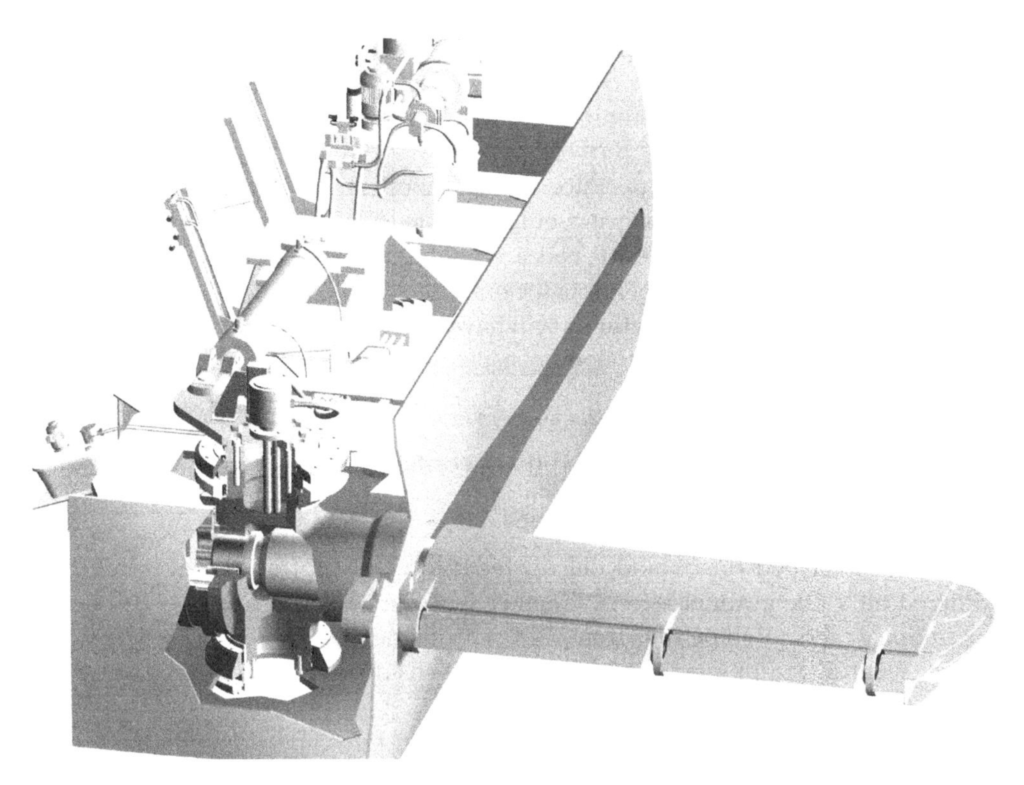

Retractable active fin stabilizer can be deployed from the ship's bridge once the vessel is in open water. The fins are housed when entering narrow waterways for berthing.

Operation of stabilizing fins

The hull form of the vessel tends to influence the choice of retractable or non-retractable fins, but the majority of wide beam, box shaped Ro–Ro vessels and passenger ships seem to favour the retractable fins. These can be 'housed' in the stowed position when in enclosed waters like harbours or when manoeuvring in narrow approach channels.

Fins are generally quite compact, having an area of between $1.2\,m^2$ to as much as $20\,m^2$, for the larger vessel. They are usually deployed by hydraulic operation with electric transmission from a bridge control station. Fins can be fitted retrospectively but are normally supplied as a pre-assembled modular unit. Lubrication oil tanks are positioned about 3–4 metres above the waterline with connecting pipe work to essential elements.

When operational, active fins would seem to be unrivalled with ship's speeds over 15 knots. Systems responding from sensed elements monitoring the deployed fin are caused by an actuator of either a hydraulic cylinder or rotary vane motor.

Roll motion on the vessel is detected by one or more sensor detectors (gyros) which feed an output signal to an integrator unit which, in turn, determines the energy of the ship's roll. This signal is then amplified and operates the hydraulic system turning the 'fins'. Feedback from each of the fins would indicate the amount of tilt being produced. Once the tilt signal equals the ordered required tilt angle the pumps stop.

Fin controls would include speed and displacement settings as well as sensitivity for varied sea conditions. Other controls may include a low speed cut out and alarm unit if the vessel falls below a minimum speed and/or a list correction to prevent fin operation on a natural list.

To effect stabilization, the operator would cancel any safety device set against unauthorized use. A start control would activate motors, sensors, alternator and hydraulics. Once pumps have attained a working pressure, locking pins would be removed prior to deployment of fins to the out turned and operational locked position.

Obviously different manufacturers will have their own terminology for fin operations but most will control the following aspects:

HOLD – Zero fin angle
STAND-BY – Produces zero lift, i.e. idling position
STABILIZE – Normal operating mode

The aerofoil shaped fins would deliver feedback to a control display against the ordered tilt value in tonnes/force together with fin angles at all times during the operational mode. Roll periods from 7–30 seconds can be handled, achieving with a 90 per cent roll damping, turning a 30° roll into an equivalent 3° damped roll.

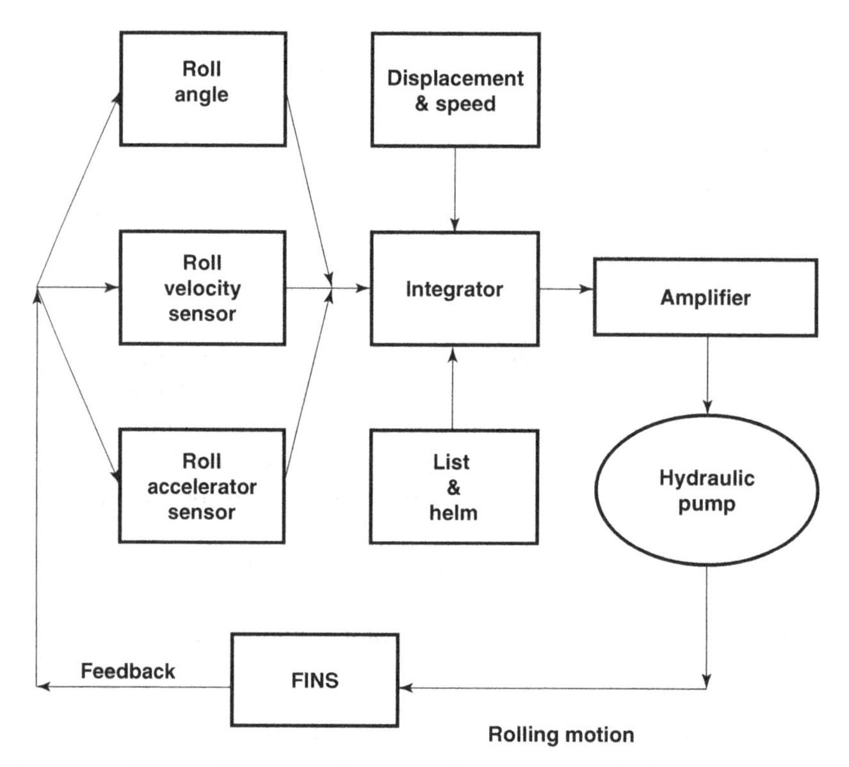

Control of fin stabilizers. Flow chart illustrates the main sensed elements which influence active fin motion.

Fixed stabilizer fins. Non-retractable trapezoidal stabilizer fin seen exposed from the hull of a vessel in dry dock. The angle of the fin can be changed by use of hydraulic cylinder action or by a rotary vane motor.

Tank systems for stabilization

Passive (internal) anti-rolling tanks

This is an internal tank system which has no moving parts other than a movement of water. The principle of the operation depends upon the movement of water lagging behind the movement of the roll of the ship by approximately 90°. This lag is achieved by adjusting the rate of water flow in the second phase of the roll. The overall effect of this delay is that the water is always flowing 'downhill' and represents 'kinetic' energy (energy of movement). The ship's roll motion provides potential energy by lifting the water volume to allow it to flow downhill. The potential energy is converted to kinetic energy which is absorbed and produces a damping action on the ship's motion.

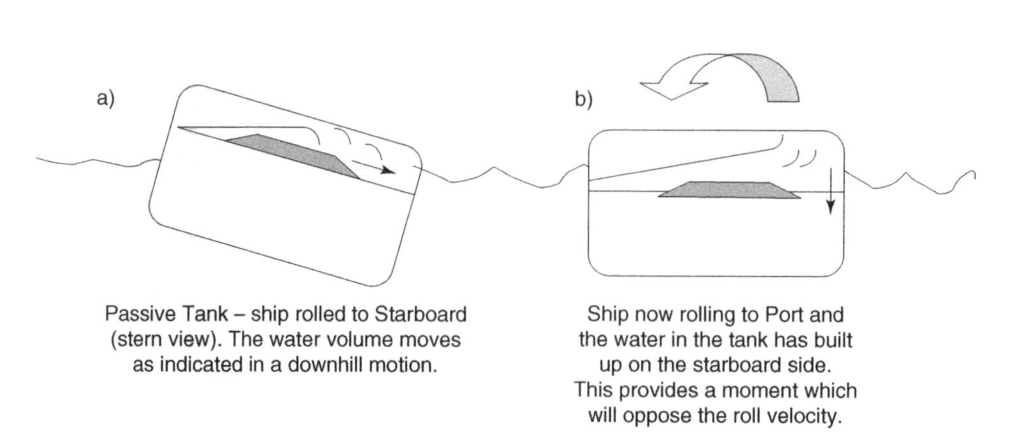

Passive Tank – ship rolled to Starboard (stern view). The water volume moves as indicated in a downhill motion.

Ship now rolling to Port and the water in the tank has built up on the starboard side. This provides a moment which will oppose the roll velocity.

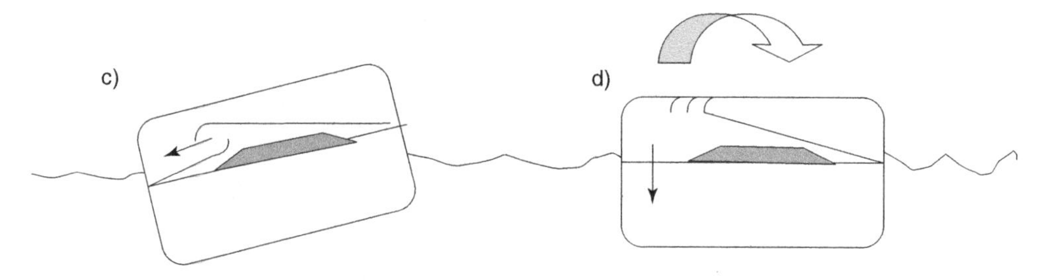

Ship now at a position of complementing the roll to port. The water is providing no moment to the ship.

Ship now rolling to starboard. The water in the tank on the port side acts to provide a moment, again, opposing the roll velocity.

Summary

This text has been compiled to expand the subject of ship handling procedures and the varied equipment associated with the task. It is not meant to provide every answer to every problem which exists in the world of manoeuvring ships. Nor can it possibly put the years of required experience on the newly qualified Master. It has been written to provide some insight to the theory, which supports an ever so practical topic for today's handlers of the world's ships.

It is a sad indictment of our training techniques that many Chief Officers gain their rank and have little opportunity to obtain 'hands on' ship handling practice. Yet as soon as that same Chief Officer is promoted to command, he is immediately expected to handle the manoeuvring of the vessel, as if he had carried out the task for years.

This element of the marine industry continues to race through change with new innovations in hardware being developed on virtually a daily basis. We saw the changes coming slowly with the advent of the Controllable Pitch Propellers. However, since that dawn, major inroads into azi-pods, thrusters, high performance rudders, new hull forms, increased speeds and improved steering concepts have changed the so-called norms of manoeuvring ships.

It is hoped the text and illustrations of this volume will go some way to help the young sea-going officer move towards the experience of the established Master.

Bibliography

Barrass, C.B. (1989) *Squatting of Ships Crossing in a Confined Channel*, Article in 'Seaways' Magazine, November.

Clark, I.C. (2005) *Ship Dynamics for Mariners*, published by The Nautical Institute.

Henson, H. (2003) *Tug Use in Port*, 2nd edition, published by The Nautical Institute.

House, D.J. (2002) *Anchor Practice: A guide to industry*, published by Witherby.

International Marine Organization (2004) *SOLAS (Safety of Life at Sea)*, Chapter V, published by IMO Publication.

OCIMF (1992) *Mooring Equipment Guidelines*, 2nd edition, published by Witherby.

Rowe, R.W. (2000) *The Ship Handlers Guide* 2nd edition, published by The Nautical Institute.

Schneekluth, H. and Bertram, V. (1998) *Ship Design for Efficiency and Economy*, published by Butterworth-Heinemann.

Watt (1970) *Vessel Performance in Confined and Restricted Channels of the St Lawrence River*, An MoT Report.

Willerton, P.F. (1980) *Basic Ship Handling for Masters, Mates and Pilots*, published by Newnes.

Williamson, P.H. (2001) *Ship Manoeuvring Principles and Pilotage*, published by Witherby.

The Maritime and Coastline Agency, Maritime Guidance Notes:
MGN 199 (M) ***Dangers of Interaction***
MGN 301 (M + F) ***Manoeuvring Information on Board Ship***
MGN 308 (M + F) ***Mooring, Towing or hauling Equipment on all vessels – Safe Installation and Safe Operation.***

Self-examiner – Questions and Answers on ship handling

Vessel manoeuvring principles

Question 1. What is the approximate position of the vessel's pivot point, when the vessel is making headway?

Answer: When the vessel is making headway the position of the ship's 'Pivot Point' is established approximately 0.25 L, from forward (where 'L', represents the ship's length).

Question 2. When a vessel undergoes ship's trials, a turning circle to port and to starboard is usually conducted. Assuming the calm conditions are the same and the vessel is fitted with a single right-hand fixed propeller, which turn would be completed tighter and quicker?

Answer: A turn to port would normally be expected to be tighter and quicker than a turn to starboard, assuming the conditions are the same.

Question 3. When operating astern propulsion, how would the ship's Master know that the vessel is actually making sternway through the water?

Answer: Observation of the water surface, from the bridge wing, would indicate that the stern wake is moving forward towards the midship's position. This indication would show the propeller position has moved aft, away from the agitated water and is actually moving the vessel astern.

Question 4. When is the rudder considered effective?

Answer: When a stream of water is passing aft of the rudder position.

Note: *The reason why the rudder is always placed amidships, when the vessel is operating astern propulsion, is that it is non-effective when water is moving forward and not passing the rudder.*

Question 5. When conducting a turning circle, which is the larger diameter scribed:

a. The tactical diameter?
b. The final diameter?

Answer: The tactical diameter is the larger of the two.

Question 6. If a vessel is required to complete a round turn when engaged operationally, what features and characteristics would affect the size and quality of the turn?

Answer: The turning ability of a vessel will be directly influenced by the following:

a. Light or loaded condition. When in a light or ballast condition the vessel will be influenced by the wind and may make considerable leeway.
b. If the vessel is trimmed by the stern, she will generally steer more easily.
c. If the vessel is not upright and carrying an angle of list the turn would take longer and turning towards the angle of list would increase the turn size. Beam of the ship will affect the turn. A narrow beam vessel will turn tighter than a broad beam vessel.
d. Additional factors affecting the turning ability include: depth of water, draught of ship, speed of vessel and the type of rudder being employed.

Question 7. What shipboard elements are under the control of Masters and Marine Pilots when involved in practical ship handling?

Answer: The control elements at the disposal of the ship handler include:

a. Engines and propulsion power.
b. Rudder(s) and steering elements.
c. Anchors.
d. Mooring ropes and lines.
e. Tugs when responding to command control.
f. Bow/stern thrust units (if fitted).

Some vessels are now fitted with Double Acting controls and, as such, would have functional ahead and astern controlling stations, inclusive of Bow Rudder configuration.

Question 8. What elements are not under control when involved in ship handling?

Answer: Clearly the weather elements during operations such as: wind, direction and force; together with state of visibility, tides and current flow, depth of water; man-made structures such as bridges; geographic obstructions like narrows and islands, as well as other traffic in the vicinity.

Question 9. When operating astern propulsion with a right hand fixed (RHF) pitch propeller the ship's stern will pay off, either to port or starboard because of the effects of transverse thrust. Which direction will the stern move when operating astern with a RHF propeller, rudder amidships?

Answer: When moving astern, rudder amidships and right-hand fixed propeller, the stern will move to port and the bow will move to starboard, because of transverse thrust effects.

Question 10. What do you understand by the term 'Headreach'?

Answer: Headreach is described by that distance a vessel will move forward over the ground, after main engines have been stopped.

Manning and station requirements

Question 1. When involved in manoeuvring the vessel, what personnel would you expect to be involved in the Bridge Team?

Answer: The Master (Team Leader), the Marine Pilot, the Officer Of the Watch, the Helmsman, the Lookout(s) and an engine room contact. Additionally, a Communications Officer and/or a radar observer.

Question 2. When taking up an aft or forward mooring station, what would the expected duties of the Deck Officer be?

Answer: The Officer in Charge of the station would be concerned with the safety of everybody aboard and be required to inspect the immediate deck area to ensure that no obstructions or hazardous situations exist. The Officer in Charge of the station would further check that all machinery is functioning correctly and that clear communications are available to the bridge. Heaving lines, stopper arrangements and any special signals are readily available for use as required. Mooring lines intended for use should be cleared from drums and flaked on deck ready for running in a safe manner.

Question 3. When picking up the Marine Pilot from a pilot launch, what preparations should be made ready at the boarding station?

Answer: Assuming that previous communications between the ship and the pilot launch have been established, the boarding station should have the pilot ladder (or pilot hoist) deployed and ready at the required operational height above the water surface. Manropes and stanchions should be rigged. Ancillary equipment such as lifebuoy and heaving lines should be readily available. Where the operation is to be carried out at night, the boarding station should be illuminated overside and in the deck area of boarding.

The Officer Of the Deck, when meeting the pilot, should inspect the rigging and securing of the ladder and ensure that communications to the bridge are operational and in good order. Where a pilot hoist is to be used, the hoist and all control functions should be checked before the pilot boards.

Note: *Where a high climb (over 9 metres) is expected, a combination rig with the accommodation ladder and pilot ladder would be arranged.*

Question 4. What international code signal flag should be displayed by the ship, once the Marine Pilot has boarded and taken the 'con'?

Answer: The ship should display 'H' (Hotel Flag) indicating there is a pilot on board.

Question 5. When a vessel is approaching an anchorage, what preparations would you expect the anchor party to carry out, before working anchors and cables?

Answer: Assuming that the Master and Chief Officer of the vessel have jointly agreed the content of the anchor plan, the Officer in Charge of the anchor party would expect to obtain power on deck and take charge of the forward mooring deck. The windlass would be inspected, oiled and greased and then turned over to ensure that the machinery was in good running order and without defects. The brake system would be checked and if found satisfactory the anchors would be placed in gear and the securings cleared away. The anchor for use would be walked back out of the hawse pipe and if it is intended to let go, it would be placed on the brake and the windlass taken out of gear. The Bridge should then be informed that the anchor is cleared away and ready to be released.

Where heavy anchors are employed, as with VLCC vessels, it is anticipated that the anchor will be walked back all the way and not let go, as is common with the smaller vessel and less heavy anchor.

Relevant lights and signals would be prepared for immediate display, once the vessel is anchored.

Question 6. A Marine Pilot is to be delivered to the vessel by helicopter. What preparations would the Heli-deck landing party carry out, prior to commencing a hoist operation?

Answer: The Heli-Deck party would prepare the hoist/landing deck area prior to the arrival of the aircraft. All preparations would be as per the ICS Guide to Helicopter Operations at Sea.

Such preparations would include the lowering of all exposed high rigging in the area of operation. The deck space and surrounding area would be cleared and any loose material secured away. A wind sock or signal flags would be displayed to provide indication of the wind direction to the pilot of the aircraft.

A designated hook handler would be equipped with rubber soled boots, rubber gloves and an insulated static hook rod, to assist the transfer. Other crew members of the deck party would be briefed on what not to do. Correct navigation signals to reflect restrictions in ability to manoeuvre would be displayed and the rescue boat would be turned out ready for emergency launch, if required.

Question 7. When weighing anchor what relevant reports are passed from the forward station to the bridge?

Answer: The direction of which way the cable 'leads' is usually indicated throughout the period of weighing anchor. The bridge would also be informed when the cable is 'up and down' and when the anchor is actually 'aweigh'. Once the anchor is clear of the water surface, confirmation that the anchor is 'sighted and clear' is usually communicated to the ship's Master. Once the anchor is stowed and secured for sea, stations would normally be stood down. The Officer in Charge of the anchor operation would be expected to report to the bridge to inform the Duty Officer and the Master of this fact and cause an entry to be made in the log book.

Question 8. When a vessel is expecting to take tugs fore and aft, what information will the Officers at these stations require?

Answer: Securing tugs to a parent vessel can be carried out by several methods and any officer under orders to secure a tug would need to ascertain:

a. Whether the tug's line or ship's line is to be employed.
b. What lead is required to secure the tug by.
c. What method of securing to the 'bitts' is required (eye or figure eight to bollards).

Question 9. When manoeuvring in close proximity to small craft, like pilot launches and tugs, what is considered a main danger and hazard?

Answer: Where small crafts of any kind are engaged with a larger, parent vessel the main danger is from the forces of 'Interaction' between the two crafts.

Question 10. Following an emergency incident at sea, like an onboard fire, what operational stations would you expect to be manned and brought to an alert status?

Answer: With most sea-going emergency incidents, the bridge team would be called in and the Engine Room would be placed on immediate stand-by. Depending on the nature of the incident then emergency parties, for damage control, rescue boat crew, a fire party or first aid party or a combination of these could expect to take up stand-by positions.

***Note**: It must be anticipated that the Master would take the 'con' and effect control of the Navigation Bridge.*

Ship manoeuvring operations

Question 1. How could a vessel turn around sharply in a river, where sea room is limited?

Answer: A vessel can achieve a tight turn by 'snubbing around' on the ship's anchor. Alternatively, depending on the amount of sea room, the vessel may be able to carry out a short round turn.

Question 2. What do you understand by the term 'Dredging the Anchor'?

Answer: This is the deliberate act of paying out an anchor, usually at short stay, with the intention of dragging the anchor on the bottom with the motion of the ship. It is often employed when approaching a berth as a means of slowing the motion of the ship's head movement.

Question 3. Why would a Baltic Moor be considered to normal berthing alongside?

Answer: A Baltic Moor is an option when the vessel is faced with an unfended, concrete quay, which could cause damage to the ship's hull plating. Alternatively, where a weak timber jetty is constructed and a heavy ship landing alongside could well demolish the flimsy jetty construction.

Note: *The Baltic Moor employs the anchor cable and a stern mooring to hold the vessel off the quay.*

Question 4. A vessel is secured alongside in Hong Kong when the Port Authority issue an 'all ships' warning of an impending Tropical Revolving Storm expected to strike the harbour area, imminent. What options are available to the Master and which option is most favoured?

Answer: In every option, the Master would terminate cargo work, batten down the vessel and consider the following options:

a. Secure the ship, let go moorings, and run for open sea.
b. Move the vessel to a storm anchorage.
c. Secure the vessel and remain alongside.

Option (a) is the best alternative provided that the vessel can get underway quickly, and clear the harbour entrance to make open waters.

The other options would include laying anchors and would run the risk of being hampered in manoeuvres. Staying alongside would also run the risk of quay damage affecting the ship.

Question 5. When carrying out a running moor, which anchor would be released first?

Answer: The sequence of letting go the right anchor, in the right order, is essential in avoiding a foul hawse situation. In the case of the running moor, the vessel should be stemming the tide and have both anchors ready for deployment. The weather anchor should be deployed first (the sleeping cable). This would be payed out to the desired scope and then the leeward anchor could be deployed second (riding cable).

Note: *When carrying out a standing moor the opposite order of anchor use is employed.*

Question 6. A vessel is experiencing heavy weather and starting to pitch heavily. Some pounding of the fore end is beginning to affect the ship. What action should the Officer Of the Watch take?

Answer: The speed of the vessel should be reduced to avoid structural and cargo damage. The Master should be kept informed of any change of the ship's speed.

Question 7. A vessel lies at anchor when another ship approaches on a collision course. How can the anchored vessel avoid the line of approach of the incoming vessel?

Answer: When threatened with impending collision, a vessel at anchor has difficulty in keeping out of the way of a potential collision. Attention to the situation can be achieved by giving five or more short and rapid blasts on the ship's whistle. However, if the approaching vessel does not respond it must be expected that the anchored ship must take whatever action she can to avoid the contact. Provided that the engines have been left on stop, with immediate readiness, the most effective action would be for the ship to steam over her own cable. Conducting an anchor operation would probably take too long and may not be that effective.

The alternative would be to give the vessel an immediate sheer by moving the rudder hard over to the side of turn. This may not allow the incoming vessel to clear completely, but it may reduce the impact to a glancing blow so limiting any damage to own ship.

***Note**: The rudder remains effective with a flow of water past it, so generating the sheer.*

Question 8. A vessel has moored with two anchors down in the form of a running moor. During the night the wind changes and causes the vessel to swing, generating a foul hawse between the two cables. What options are available to the ship's Master, in order to clear the foul?

Answer: In order to clear the foul hawse the Master's options are:

a. Try to turn the ship with engines and rudder action in opposition to the foul turns in the cables.
b. Engage a tug to push the vessel around in the opposite direction to the turns.
c. Make use of a motorized barge. Break the sleeping cable and lower the bare end of cable into the barge, drive the barge in opposition around the riding cable, then rejoin the cable together.
d. Break the sleeping cable on deck and dip the bare end half a turn at a time, under and over the riding cable to take out the foul turns. Rejoin the cables once the foul is clear.

Question 9. A vessel is approaching a bend in a river and hears a sound signal of one prolonged blast followed by two short blasts at intervals of two minutes. What action should be taken by the vessel approaching the bend?

Answer: The sound signal should be recognized as a fog signal of another vessel. As such a poor state of visibility exists around the bend in the river, the ship should make immediate preparations for entering a fog bank. Bridge actions and preparations would include: reducing the ship's speed, commencing sounding fog signals, posting extra lookouts, switching on navigation lights, informing the Master and relevant departments.

***Note**: With the vessel already in position in the river, it is probable that the Master and a full bridge team may already be in situ, and that engines may already be on stand-by for manoeuvring speed.*

Question 10. When manoeuvring the vessel, at what times would the manoeuvring light be required to supplement the use of the ship's whistle?

Answer: The light can be used to supplement the sound signal at any time deemed necessary and many ships are often fitted with a manoeuvring light that activates automatically with the operation of the ship's whistle. However, the range of the whistle is limited (average two miles depending on weather), so where the range of a target vessel if greater than the two miles, it may be considered prudent to supplement sound signals by the light signal.

Mooring operations

Question 1. How would you moor the vessel with reduced swinging room in a river or canal?

Answer: A ship can achieve reduced swinging room by use of a running or standing moor operation.

Question 2: An A.C. 14 anchor is considered a high holding power anchor over and above a converted stockless anchor.

a. What is the holding power difference between the two anchors?
b. Why is the difference in holding power generated?

Answer:

a. An A.C. 14 anchor will have approximately ten times its own weight in holding power, while a converted stockless anchor will have approximately four times its own weight.
b. The increased holding power on an A.C. 14 is achieved by its prefabricated construction, shape and fluke surface, as compared with the cast construction of a conventional stockless anchor.

Question 3. What factors would a ship's Master consider when deciding on the amount of scope to use, when bringing the vessel to an anchorage, with a single anchor?

Answer: It is assumed that prior to approaching the anchorage, an anchor plan for the operation has been completed. The consideration of the amount of chain cable to use is influenced by the following features:

a. The depth of water in the intended anchorage.
b. The draught of the vessel.
c. The range of tide expected inside the anchorage area.
d. The prevailing weather and expected weather.
e. The length of time for which the vessel intends to be anchored.
f. The holding ground for the anchor position.
g. The holding power and type of anchor being employed.
h. The rate of any current or tidal stream.

Note: *It is always considered that it is not just the anchor but the amount and lay of cable that keeps the anchored vessel reasonably secure and protects from dragging.*

Question 4. When mooring to buoys, how would the mooring lines be secured to the ring of the buoy?

Answer: There are several methods that can be used to secure soft eye mooring lines to a buoy ring, namely:

a. Use of a mooring shackle from the eye directly onto the buoy ring.
b. By means of a toggle lashed under the bight and over the rope eye.
c. By use of a manila lashing securing both sides of the eye beneath the rope bight.

Question 5. When would you expect a vessel to carry out an open moor operation?

Answer: An open moor is employed in non-tidal waters where the additional strength of a second anchor is required.

Question 6. What are the advantages and disadvantages of a Mediterranean Moor?

Answer: Advantages of the Mediterranean Moor are:

a. More vessels can be secured to the quay, stern to, when quay space is restricted.
b. Cargo ships can work both port and starboard at the same time into barges.
c. Ro–Ro vessels can carry out stern load/discharge by use of a stern ramp.
d. Tanker vessels may discharge aft, via a stern manifold.

Disadvantages of the Mediterranean Moor:

a. The vessel is left exposed bow out, into open water.
b. The ship is denied the use of shore side cranes and must use own cargo gear.
c. Going ashore requires the use of a small boat.

Question 7. What angle off the bow would you expect to lay anchors when conducting a Mediterranean Moor?

Answer: Ideally, each anchor should be laid so that the cable is approximately 15° either side of the fore and aft line.

Question 8. When departing from a position, following a running moor operation, which anchor cable would be recovered first?

Answer: The vessel would expect to drop astern and recover the 'sleeping cable' first; then, with the use of engines, bring the vessel forward to recover the 'riding cable' last.

Question 9. What is the main disadvantage of any mooring with two anchors deployed?

Answer: The ship always runs the risk of acquiring a 'foul hawse' when two anchors are deployed, especially if the wind direction changes and this goes unnoticed.

Question 10. A VLCC vessel with 20 tonne anchors is ordered to go to anchor by the Port Authority. What do you consider is the main difference in the anchoring methods employed for the large vessel, compared with the smaller ship of a more conventional size?

Answer: A large vessel with heavy anchors would generally not contemplate 'Letting the Anchor Go', but walk the anchor back all the way under the power of the windlass. Such vessels would also probably require a deep water anchorage as a main consideration and the speed of approach would need to be considered in conjunction with the prevailing weather conditions. Ample time would be required to slow the vessel down prior to approaching the anchorage.

Interaction theories

Question 1. When a vessel enters into shallow water from deep water, what considerations should be noted by the Watch Officer?

Answer: The steering of the vessel may be directly affected by the changing depth. The ship may also experience an increased level of 'squat' and effectively experience a seeming loss of under keel clearance (UKC).

Question 2. How can the effects of squat be practically reduced?

Answer: It is generally agreed that the amount of squat experienced by a ship is directly related to the speed^2 of the vessel. Therefore, if the speed is reduced the effects of squat will also be reduced.

Question 3. A vessel is moving through a canal which has other vessels secured alongside. As the vessel passes the secured ships they start to range on their moorings. What is the cause of this movement and what corrective action can be taken?

Answer: The passing ship is causing interactive forces which affect the other vessels tied up, either because the other ships' moorings are slack and/or the speed of the moving vessel is too fast, causing a backwash of water movement.

Any vessel secured with slack moorings reflects poor ship keeping and they should be advised by the Marine Pilot/Port Authority. Any vessel passing another should note the condition of the moorings and note slack moorings in their log book, with the name of the ship. In any event, own ship's speed should be dramatically reduced when passing stationary vessels.

Question 4. Briefly explain 'Bank Cushion Effect'?

Answer: When a vessel is in close proximity to a bank, a pressure cushion builds between the bank and the ship's hull. This external pressure influences the bow angle away from the bank, outward. The danger here is that this unexpected outward turn may bring the vessel into close proximity of another ship.

Question 5. When the vessel is lying to a single anchor, where is the pivot point?

Answer: The vessel will pivot about the hawse pipe position in the bows.

Question 6. If a tug was scheduled to make fast with a large parent vessel in the forward position, what would the tug master be concerned with on his approach?

Answer: Ideally, the tug should take up station well in advance of the vessel's approach. By making the rendezvous in advance, the tug can await the approach of the larger vessel without involvement of interactive forces in the region of the ship's shoulder and under the flare of the bow.

Any tug master would expect to be aware of the pressure force that can be generated by two vessels alongside each other. A tug approaching parallel to the fine lines of the vessel can expect to generate a maximum interactive force at a position off the ship's shoulder. This can be counteracted by opposing helm orders. However, if the tug is left carrying that helm while making headway, she may run the risk of taking a sheer across the bows of the parent vessel, when the pressure drops away under the flare of the bow.

Question 7. When two vessels are passing from opposing directions in a narrow channel, the dangers from interaction should be realized. What reaction would you expect the two passing vessels to make, if no counter action is taken?

Answer: As the bows of the two vessels draw opposite to each other it must be anticipated that the bows will deflect outwards by the water pressure developed between the two hulls. When the bows move outward the sterns may be 'sucked in' together with a resulting collision to the stern parts.

Question 8. A vessel is overtaking another larger vessel in a narrow channel. If manoeuvring so close, that interaction forces are allowed to become involved, what reaction would you expect to occur between the two ships?

Answer: One would expect the bow of the overtaking vessel to be pushed outward while the sterns of the two vessels could be drawn together with possible collision in the stern area occurring.

Question 9. A vessel moving down a canal causes a blockage factor by its underwater volume in the gut of the canal. What would be the ongoing concerns of the ship's Master?

Answer: In narrow waters the ship will experience less underkeel clearance causing less buoyancy forces affecting the hull. The results of this could be that the vessel acquires more sinkage and may take on increased values of squat. The risk of grounding is increased and the ship may 'smell the bottom'. Adverse effects could also be noticeable when the vessel turns and heels over into a turn. The turn of the bilge could make contact with the ground during a turn, if the speed of manoeuvring is too excessive, causing an increased angle of heel.

Question 10. A parent vessel is manoeuvring with the aid of tugs. The danger from interaction between the larger vessel and the smaller tug is realized by all parties. How can the tug-master reduce the risk of the tug being girted?

Answer: Excessive or sudden movement of the parent vessel, when operating with tugs secured, could cause the towline to lead at right angles to the fore and aft line of the tug, so generating a capsize motion from near the midship's position on the tug.

Such a motion can be changed to a turning motion by use of a 'Gob' (gog) rope, causing the tug to slew, rather than heel towards a capsize.

Hardware

Question 1. When is it considered the best time to inspect anchors and cables and the chain locker?

Answer: The ideal time to carry out an inspection of anchors and cables and the chain locker, is when the vessel is in dry dock. All the cables can be removed from the locker safely to allow a detailed and visual inspection of the locker. The cables can be ranged in the dock for close inspection and the anchor can be fully exposed for easy inspection from the floor of the dock.

Question 2. What are the advantages and disadvantages of a Controllable Pitch Propeller (CPP), as compared with a righthand fixed propeller (RHF)?

Answer: Advantages of CPP over RHF:

a. More immediate and improved bridge control with CPP.
b. Vessel can be stopped without stopping main machinery.
c. Shaft alternators can be employed saving auxiliary machinery fuel.
d. Improved ship handling procedures can allow manoeuvres without the need to engage tugs, making reduced operational costs.
e. Easy to change damage blades (spare blades easy to carry).

Disadvantages of CPP over RHF:

a. Expensive to install, especially retrospectively.
b. Creep effects may occur without close monitoring.

c. Increased maintenance required.
d. Double station controls required for Bridge and Engine Control Room. Requires additional redundancy in sensors, monitors and similar hardware.
e. More moving parts and more chance of malfunction.

Question 3. What is the common advantage of modern rudders fitted with flaps, rotors and developed as high lift like the 'Schilling' rudder?

Answer: Improved ship handling performance, faster operations and greatly reduced turning circle ability.

Question 4. How often must the emergency steering gear be operated from the auxiliary steering position?

Answer: At least once every three months. A record of this operation must be kept in the Official Log Book.

Question 5. How are corrosive effects controlled in the region of rudders and propellers constructed with dissimilar metals to the hull?

Answer: Most vessels employ sacrificial anodes secured to the effected areas. Additionally, Cathodic protection is used separately or alongside anodes.

Question 6. Where a vessel is experiencing heavy pitching motions, there is a risk of the stern and propellers breaking clear of the water surface. What control element reduces the risk of screw race?

Answer: The main engine machinery will be fitted with a 'Governor' control.

Question 7. How would you turn a twin screw vessel to starboard, fitted with outward turning propellers, in reduced sea room?

Answer: The turn could be executed by going Full Astern on the starboard engine while going Full Ahead on the port engine, rudder amidships.

Question 8. What is the associated danger of working with tugs fitted with 360° rotatable thrust units or rotating ducted propellers?

Answer: The Officer on station will probably not know the type of propulsion fitted to the tug. The danger exists with towlines and moorings in the water which may become fouled in the directional propulsion units when propellers are turning to suck the ropes in, or push them away into own ship's propellers.

Question 9. What is the difference between a Balanced Rudder and a Semi-Balanced Rudder?

Answer: A Balanced Rudder will be constructed with a 25–30 per cent of its plate area, forward of its turning axis. A Semi-Balanced Rudder will only have approximately 20 per cent of its plate area forward of the turning axis.

Question 10. When testing the ship's steering gear, prior to sailing, the rudder is turned hard to starboard, then hard to port. How will the inspecting officer know if the steering systems are all working correctly?

Answer: Once the steering motors have been switched on, a tell-tale monitor on the bridge will indicate the status of each set of motors. As the wheel is turned, the helm indicator and the rudder indicator should show the amount of rudder movement caused respective to the amount of helm being applied. Telemotor systems will also have oil pressure gauges at the helm position which would indicate the pressure levels held at the hard over positions. Should these levels fluctuate then a loss in pressure levels is being experienced and would reflect a possible defect.

Emergency manoeuvres

Question 1. A vessel loses her rudder when off a lee shore. What options are available to the ship's Master?

Answer: The Master would immediately go to a Not Under Command status and display the appropriate signals. The nature of any emergency communication that may have to be transmitted will depend on the proximity of the shoreline. If the situation is life threatening, then a distress MAYDAY signal would be required. Where the situation may be relieved and could be delayed to resolve any immediate danger, an URGENCY signal may be a suitable alternative. It should be realized that Maritime Authorities would rather be informed sooner than later, in order to effect immediate contingency planning for the incident.

Immediate actions will depend on the capabilities and resources within the type of vessel involved. For example, if a twin screw vessel is left without normal steerage, then the ship could initially steer by use of engines, adjusting the revolutions on one side or the other in order to turn away from the danger.

Emergency use of the ship's anchor(s) could also be a suitable delaying tactic – by reducing the drift rate of the vessel, towards a potential hazard. However, such use of an anchor may be in deeper water than one would normally expect. If such was the case, the anchor should be walked back all the way; the objective of this being to hold the vessel off the shoreline until an ocean going tug could be engaged.

***Note**: With the increasing size of ship builds previous remedies, like the rigging of a jury rudder, is seemingly no longer practical (small vessels are an exception). However, where a ship can consider exceptional circumstances, it may be possible to employ drag weights to produce a drogue effect from either each bow position or from the vessel's quarters. It should be realized that with limited crew numbers and possibly no own ship's lifting gear (as with large bulk carriers), deployment of heavy materials might be nearly impossible.*

If the rudder is lost, then the situation cannot be resolved to return full control to the ship without taking the vessel to a dry dock. Where the rudder is lost, the positive end solution is to engage a tug (or tugs) to manoeuvre the vessel towards a repair facility.

Question 2. A vessel experiences an on-board fire. Following the fire alarm being raised the Master takes station on the navigation bridge and takes the 'con'. What manoeuvring tactics could be employed to be beneficial to the fire fighting operation?

Answer: A recommended method of fire fighting is to starve the affected area of oxygen. By turning the vessel stern to wind and adjusting the ship's speed, provided available sea room is available, this action could expect to reduce the draught and hence the oxygen content within the vessel and around the fire scenario.

An exception to this action may be appropriate where the fire is generating a high volume of smoke and a positive draught is required to clear smoke from the immediate area.

Question 3. A man is suddenly lost overboard, what type of manoeuvres would be appropriate for the Officer Of the Watch to take, in order to effect recovery of the man in the water?

Answer: The IAMSAR Manual, Volume III, recommends suitable manoeuvres to recover a man overboard. These are:

a. A Williamson Turn.
b. A Single Delayed Turn.
c. A Scharnurst Turn.
d. A Double Elliptical Turn.

Question 4. Following a manoeuvring turn to effect recovery of a man overboard, the vessel returns to the 'Datum'. No sign of the man overboard is seen, what type of search pattern should be commenced?

Answer: It is a legal requirement that the Master engages in a search for the missing man. The type of search pattern chosen would be at the discretion of the Master and be dependent on the prevailing weather conditions. Where the 'Datum' is known to be reliable, then a 'Sector Search' would probably be adopted, with a small track space.

Question 5. A ship requires to make a medical evacuation of a crew member and a rendezvous with a helicopter in the Irish Sea is scheduled. What conditions would the ship's Master want to impose and how would he want to have the ship heading, in relationship to the wind direction, when engaging with the aircraft?

Answer: The Master would need to ensure that the operation was conducted safely, in open water, clear of other traffic and navigation obstructions. There should be adequate under keel clearance throughout the period of engagement with the aircraft. A Helideck landing Officer should be appointed and the deck area should be cleared of loose objects and any high obstructions that could compromise the rotors of the helicopter.

Correct navigation signals should be displayed, as for a vessel restricted in ability to manoeuvre. A wind sock or other suitable indicator (flags) should be shown to inform the aircraft pilot of wind direction.

The Master would take the 'con' of the vessel and make a course with the wind direction approximately 30° off the Port Bow. This permits the aircraft to hold station, presenting its starboard (winch) side, to the port side of the vessel, while heading into the general wind direction to achieve positive directional control.

Question 6. A high-sided car carrier responds to a small boat sinking and in distress. Rough sea conditions prevail in an estimated Gale Force '8'. How could the ship's Master effect recovery of the boat's survivors, while minimizing the risk to his own crew?

Answer: The rough sea conditions would make it foolhardy to attempt to launch own ship's rescue boat. An alternative strategy could be to secure own ship's lifeboat to the boat falls and prepare to lower the boat towards the surface, with the intention of using the lifeboat as an elevator to recover survivors from the surface.

Ideally, the parent vessel should create a 'lee' for the approach of the small boat. Keep own ship's lifeboat fully secured on the falls and on approach of the survivors' craft, lower own ship's lifeboat to the water surface, to enable survivor transfer from one craft into the other. Once all survivors have boarded the ship's boat on the falls, hoist the lifeboat clear of the water and disembark at the ship's embarkation deck prior to securing own lifeboat.

Note: *Retaining the lee for the transfer to take place from one craft to another will be a difficult task for the ship's Master on the con. Too much wind on the bow during the transfer could cause the vessel to set down over the survival craft.*

Question 7. Following a collision a damaged vessel is forced to beach in order to prevent the vessel from sinking. On the approach to the ground it is realized that the ship is beaching on a rising tide. What are the dangers and concerns for the ship's Master?

Answer: Depending on where the ship is damaged the Master's concern would be to take the beach in order to save a total constructive loss. Where the operation is on a rising tide, the possibility of the ship accidentally re-floating itself is a real one and the Master would want to take all measures to retain the ship in position on the beach.

The idea of beaching the ship is with the view that it can be temporarily repaired and caused to be re-floated at a later date. To this end, the Master would probably order both anchors to be deployed once the ship reaches the beached position. This would effectively reduce the risk of re-floating and the damaged vessel prevented from dropping astern into deep water.

Note: *There is clearly a case to be made for driving the vessel further on to the beach, in order to prevent the ship re-floating in an uncontrolled manner where the loss of the vessel may occur. Prudent use of ballast could also help to retain the ship's beached position.*

Question 8. A vessel is approaching a port when a vessel aground is sighted at two compass points off the starboard bow. What would be the expected actions of the Officer Of the Watch (OOW)?

Answer: Assuming that the OOW is alone on the bridge and acting as the Master's representative, it must be anticipated that he or she would stop their vessel immediately taking all way of the ship. The Master should be advised as soon as possible of the situation. The OOW would carry out a chart assessment, to include their ship's own position and the position of the vessel aground. The OOW should also assess the area and proximity of the shoal. The echo sounder should be switched on and the underkeel clearance noted. Communication with the aground vessel should be made, following station Identification. Information of the time and date of grounding and the vessel's draught should be requested from the Officer in Charge of the vessel aground.

Once the Master takes the 'con' he would expect the OOW to report all relevant facts concerning the situation.

Question 9. A vessel strikes an underwater object while on route from Dakar to Cape Town. What following actions would be expected of the ship's Master?

Answer: A Master would be concerned about the watertight integrity of his own ship and would probably order the Chief Officer to carry out a damage assessment, to include a full set of internal tank soundings.

The position of striking the object should be noted in the Log Book, together with the time of the occurrence. These details should be transmitted to the Marine Authority, and also reported to the Marine Accident Investigation Branch.

Question 10. A vessel on route to Montreal, via the North Atlantic, experiences a high level of ice accretion. What are the dangers of this and what actions can be carried out to limit the ice build-up?

Answer: The real danger of ice accretion is from the added top weight to the vessel which could directly affect the positive stability of the ship. Where possible, the Master should alter course to more temperate latitudes and reduce speed to counter any wind chill factor.

The crew should be designated to clear ice formations from the upper structures by use of steam hoses or with axes and shovels. Crew members should be adequately protected when employed in this task and the work should be covered by a risk assessment.

Miscellaneous

Question 1. When securing the stern mooring wire to the chain, for use in a Baltic Moor, why would the wire be secured to the 'Ganger Length' rather than the Anchor Crown 'D' Shackle?

Answer: When departing the berth, the wire would be easier to release from the ganger length on deck than from the Anchor Shackle, which may be stowed well inside the hawse pipe.

Question 2. In what order of passing moorings to mooring buoys, would you expect to pass the slip wire?

Answer: Once the vessel is secured to the mooring buoy by soft eye mooring ropes, it would be normal practice to send the slip wire last. Use of a slip wire is notably the last line out and the last line in, when departing.

Question 3. What preparations would you make if on station and ordered to run a slip wire?

Answer: The eye of the slip wire must be seized and reduced in order for it to pass through the ring of the buoy. When running the slip wire to the mooring boat it would be necessary to pass a messenger line with it. Once the reduced eye has been passed through the buoy ring, the messenger can be secured to the slip wire eye so that it can be recovered as a bight on board. Once retained on board, both parts of the slip wire can be secured to the bitts.

Question 4. When weighing anchor, a wire cable is found to be fouled over the fluke of the anchor. What action would you expect the Officer in Charge of the anchor party to do?

Answer: The situation should be reported to the Master and the Navigation Bridge. The possibility of letting the anchor go again, may be successful in bouncing the cable off the fluke. Alternatively, the wire should be stayed off and the anchor walked back to clear the fluke angle. This would allow the anchor to be drawn home and the wire could be cast off.

***Note**: A boatswain's chair operation overside could allow the troublesome wire to be tied off to clear the anchor. However, all precautions as per the code of safe working practice must be complied with in rigging the chair. Also, a risk assessment and permit to work overside would need to be carried out prior to the operation taking place.*

Question 5. A ship becomes 'beset' in pack ice and requires the services of an ice breaking vessel, to break free. Where could the ship obtain information about the ice breaker services available?

Answer: Of the official publications carried on board, the Sailing Directions (Pilot Books) would be expected to provide details of ice breaker services available to ships navigating in the region. Additional general information on ice could also be obtained from the Mariners Handbook. Communication details can be sought from the Admiralty List of Radio Signals.

Question 6. A vessel is to enter a dock from a tidal river and no tugs are available. What would be a suitable manoeuvre to dock the ship safely?

Answer: The ship should stem the tide and go alongside on the dock wall below the dock entrance. By use of carrying up the mooring ropes the vessel can be warped around the knuckle entrance into the dock area.

Note: *A pudding fender might be considered useful as the ship turns on the knuckle to enter the dock area.*

Question 7. What is considered good 'holding ground' for a ship going to anchor?

Answer: Mud or clay are the better types of holding ground as they tend to grip and hold the anchor better than marsh, ooze or rock (considered bad holding ground).

Question 8. In accord with the COLREGS (Regulation 35), a vessel aground in restricted visibility may make an appropriate whistle signal. What is considered an appropriate whistle signal?

Answer: Use of the international code letters 'U' or 'L' may be appropriate.

Question 9. A vessel is required to carry out a swing in order to check the magnetic compass. What conditions are required in order to complete this manoeuvre safely?

Answer: Whenever a compass swing is required it would be expected that the manoeuvre would be carried out in an area free of traffic and clear of magnetic anomalies. A fixed landmark would be used, or alternatively the sun could be employed, to take a set of bearings from the swing position. The swing should take place with the vessel upright and with adequate underkeel clearance for the ship throughout. No electrical influences should be near the compass site and no other ships should be within three cables distance.

Question 10. Why is it necessary to take an azimuth or amplitude to check the magnetic compass on every occasion of a major alteration of ship's course?

Answer: A compass check is a method of obtaining the Deviation of the compass, which changes with the direction of the ship's head. The algebraic sum of deviation and variation determines the compass error. In the event of malfunction of the gyro compass, the ship would have to navigate by means of the most important instrument aboard, namely the magnetic compass. In order to do this successfully, it would always need to be able to apply the compass error (variation being noted and obtained from the navigational chart, respective to the ship's position).

Index

9780750685306